The VFUU Price of Oil charts the story of oil, fro
its transformation of every aspect of our lives, and concludes with a compelling narrative of why we now need to move beyond oil once and for all. I was engaged, entertained and enlightened by this unique approach to the subject.

Ben Goldsmith | CEO of Menhaden Capital Plc, founder of WHEB Asset Management | http://www.menhaden.com/

Oil has served us well and stimulated the biggest growth and wealth creation in human history. *The VFUU Price of Oil* explains the role it has played and why it is time to wean ourselves off our addiction to oil.

Paul Polman | Chief Executive Officer, Unilever | www.unilever.co.uk

Michelle Spaul provides a broad and very readable introduction to the world of oil. Starting from a perspective of the societal advances that have been enabled by oil she looks at how its price influences nearly all aspects of human activity. She then moves on to look at the costs, on the environment and climate and their ramifications, and concludes that we should all be looking to remove oil from our lives.

Professor Joanna D. Haigh CBE FRS | Co-Director, Grantham Institute – Climate Change and the Environment | www.imperial.ac.uk/grantham/

the

volatile fickle unpredictable undependable

PRICE OF OIL

by

Michelle Spaul

Published by Delta Swan Limited

Copyright @ Michelle Spaul 2016

First published in 2016

Print book ISBN 9781849149273

Printed Lightning Source International

www.michellespaul.wordpress.com

With Kind Permission from...

The following organisations and individuals have kindly granted permission for the use of quotes:

What's it oil about: *"Ernie..."* Contains Parliamentary information licenced under the Open Parliament Licence v3.0 http://www.parliament.uk/site-information/copyright/open-parliament-licence/
http://hansard.millbanksystems.com/commons/1944/jun/21/employ ment-policy

Here, there and everywhere: *"see nuclear ships sooner..."* from 'Consortium explores nuclear power for ships' – Marine Log - Nov 2010 http://www.marinelog.com/index.php?option=com_k2&view=item&i d=186:consortium-explores-nuclear-power-for-ships&Itemid=227

Plastic dolls: *"beakers, buckles, combs..."* The Worshipful Company of Horners
http://www.horners.org.uk/cms.php?cms_id=47

Taking all my energy: *"The UK model..."* from 'All hands on deck for an electricity local generation', a letter to the Financial Times from Alistair Buchanan
http://www.ft.com/cms/s/0/d4200334-95bc-11e4-a390-00144feabdc0.html?siteedition=uk

Taking oil my energy: *"sited with very little consideration..."* from 'Common Concerns about wind power' by the Centre for Sustainable Energy - May 2011
http://www.cse.org.uk/downloads/file/common_concerns_about_wi nd_power.pdf

Energy: *"Economic and social development in sub Saharan Africa..."* © OECD/IEA 2014, Energy sector is key to powering prosperity in sub-Saharan Africa, IEA Publishing. Licence: http://www.iea.org/t&c/termsandconditions
https://www.iea.org/newsroomandevents/pressreleases/2014/october /energy-sector-is-key-to-powering-prosperity-in-sub-saharan-africa.html

It's the economy, stupid: *"The IEA (International Energy Agency) is…"* © OECD/IEA undated, About us, IEA Publishing. Licence: http://www.iea.org/t&c/termsandconditions http://www.iea.org/aboutus/

It's the economy, stupid: *"The mission of the Organisation for Economic Co-operation and Development…"* from the OECD's website http://www.oecd.org/about/

It's the economy, stupid: *"The Overseas Development Institute (ODI)…"* from the ODI's website http://www.odi.org/about

It's the economy, stupid: *"The International Monetary Fund (IMF)…"* from the IMF's website http://www.imf.org/external/about.htm

Peak Oil: *"we have to leave oil…"* from Warning: Oil supplies are running out fast [interview with Fatih Birol], The Independent - Aug 2009 http://www.independent.co.uk/news/science/warning-oil-supplies-are-running-out-fast-1766585.html

Appendix Six: *"Scientific consensus is…"* from 'Scientific consensus' (without alteration, transformation or being built upon, under Creative Commons Attribution-ShareAlike 3.0 Unported License http://creativecommons.org/licenses/by-sa/3.0/) Wikipedia - Sep 2015 https://en.wikipedia.org/wiki/Scientific_consensus

To oil, or not to oil: *"1. Increasing energy efficiency"* © OECD/IEA 2015, World Energy Outlook Special Report - Energy and Climate Change, IEA Publishing. Licence: http://www.iea.org/t&c/termsandconditions https://www.iea.org/newsroomandevents/pressreleases/2015/june/iea-sets-out-pillars-for-success-at-cop21.html

Appendix eight: *"Ofgem is…"* from Who we are, the Ofgem website https://www.ofgem.gov.uk/about-us/who-we-are

Contents

Figures

Preface

Money for oiled rope

MANY MOONS AGO, I CLEANED OUT MY GARAGE and found some tow rope. It was made of blue plastic. I started to wonder when I last used a tow rope or any other rope for that matter. I asked myself when the rough, fibrous ropes of my childhood had given way to nylon. A little research led to more questions and a few stark realisations; before long, I was writing this book.

I found rope to be an excellent example of the way oil has replaced natural fibres and the energy of animals and people, and what life without affordable oil might be like. I would like to tell you more about rope in this short preface before getting into the details of *The Volatile, Fickle, Unpredictable, Undependable Price of Oil.*

The history of rope

As humanity developed, rope became a valuable tool. Archaeological evidence, found in fossilised remains and cave art, tells us human beings have used rope for millennia. At first, we tamed animals, gathered food[1] and moved heavy objects like the Easter Island moai. As we grew more sophisticated, so did our use of rope.

---

1 Eight thousand years ago our ancestors painted images of people collecting honey while dangling from ropes.

Rope was essential in building the monuments to the Pharaohs and their families. Around 2540 BC Pharaoh Khufu (Cheops in Greek) ordered the Great Pyramid at Giza to be ready for his death. Architects and slaves spent 20 years building the edifice out of stone blocks: granite for the internal structure and limestone for the façade. Thousands of slaves dragged the blocks into place with rope. During the construction period, they moved over 800 tons[2] of material every day. Historians and engineers suggest the Egyptians had many building methods. Most involve dragging, lifting or rolling the stones; each method needs energy (for pushing and pulling) and something to hold the rocks steady. In both regards, rope had no peer for millennia. Okay – not for the pushing bit.

Rope also had a role in building empires. Khufu's wealth came from the Mediterranean Sea and Arabia and, in preparation for the afterlife, he had two ships buried outside his Great Pyramid. A few years after Khufu made his final journey, Egyptian sailors stored rope in a hand-hewn cave in the coastal settlement and port of Mersa Gawasis. In 2011, archaeologists found these 4,000-year-old papyrus ropes. They look like their modern descendants and are coiled and stored in the same way.[3]

Thousands of years later, sailors still relied on rope to explore the world, find wealth and create empires. Rope for the Royal Navy had to measure 1,000 feet (305 metres) without joins. It needed enough strength to hoist and hold sails, and to keep ships at anchor. You might think the need for rope would disappear quickly after steam power arrived in 1736, but wind power stayed around until the early nineteenth century.[4]

History of rope's manufacture

Back in the day, rope makers were respected artisans. They plied their trade in every town and city until the industrial revolution. Road signs show the reach of the industry: from the City of London (Ropemaker Street) to Hamburg's Reeperbahn, with dozens of Rope Streets, Roper Roads, Rope Walks and Rope Walk Courts in between. Understanding

---

2 You will find tons – the US/ Imperial measure – and tonnes – the metric measure – throughout this book. The measure reflects the source of the data; in this case it is old.

3 String is even older, and has been found among Neanderthal artefacts.

4 Steam gave way to diesel in the second half of the 20th century. Steam drove the RMS Titanic, RMS Queen Elizabeth and RMS Queen Elizabeth II. Cunard converted the latter to diesel in 1986.

that the best rope uses its form to provide strength, rope makers developed methods to make their product as reliable as possible.

Every method of making rope braids or twists fibres to increase their length and strength. When we pull on a rope, the friction between the fibres stops them slipping; when rope holds a weight, every fibre pulls against its neighbours, trying to break free. If they succeed, the results can be disastrous.

Ancient Egyptians made the first rope-making machines.[5] They used dowels to spin lengths of fibre into strands, which they then spun together. Medieval Europeans developed the ropewalk in the 12th century. The new method produced more consistent rope than older techniques and employed less labour, so the resulting ropes were stronger and cheaper. The ropewalk encompassed a large, covered area, or even an entire street, with hooks at each end. The rope maker laid out fibres and twisted them with the hooks.[6]

During the Industrial Revolution, machinery took the place of people and increased productivity. One of the architects of the transformation was Richard Arkwright. He designed and built water- and steam-powered mills to replace traditional cloth manufacturing (the cottage industry).[7] His first mill still stands in Cromford, Derbyshire. The Industrial Revolution changed manufacturing forever and gentlemen inventors crowded round to take advantage of and contribute to advancements in science and engineering. Edmund Cartwright, a cleric, was one. He developed the power loom, first patented in 1784 and, in 1792, he registered a patent for a mechanical rope-making machine. His Cordelier quickly displaced traditional ropewalks. Modern rope-making machines have their basis in Cartwright's invention.

Our ancestors made ropes from natural fibres, eg vines, hemp, manila, cotton, abaca and straw. Now we make them from nylon (which also makes tights), polyester (bottles), polyethene (plastic bags), acrylics (jumpers and other fabrics) and polypropylene (kettles and other

---

5 In the Sources, you will find a link to a picture of rope making in Ancient Egypt.

6 Two UK rope makers still make rope in the traditional way: The Master Ropemakers in Chatham Dockyard and the Outhwaite Ropemakers in Hawes, Yorkshire.

7 While Arkwright had the first commercial success with a watermill, Christopher Polhem built the first water-powered factory in Sweden in 1700. The latter also invented the padlock.

domestic appliances). We make all these materials from oil. Modern machines spin ropes of various lengths and diameters for many applications. You can buy rope made of natural fibres, but as it is more expensive and less strong for its weight than synthetic line, most people do not.

Our dependency on oil

Oil has many more uses than providing the energy and raw materials to make rope, as we will explore in *The Volatile, Fickle, Unpredictable, Undependable Price of Oil*. First, let's compare two similar engineering feats:

- In 1586 Pope Sixtus V ordered the engineer Domenico Fontana to move an obelisk from the old Roman circus to the centre of St Peter's Square in the Vatican City.

- In 1999 relocation saved the Belle Tout Lighthouse at Beachy Head from a watery grave.

For the obelisk, the 400-metre move absorbed the strength of 900 men and 140 horses, who used 44 winches to heave on 40,000 tons of iron bar and hemp rope. They took five months to move the solid block of stone and its base: a weight of 330 tons.

The lighthouse weighs 600 tonnes and its owners moved it 17 metres in two days.

The key difference between the two moves was oil. Oil fuelled the vehicles, powered the hydraulic pumps and made the plastics housing the computers. Without oil, the owners of Belle Tout would have needed strong backs hauling on rope. Except today, rope uses oil too.

The consequences of dependency

In *The Volatile, Fickle, Unpredictable, Undependable Price of Oil* we will learn that we cannot depend on stable, economically viable oil prices – and that means:

- Our ports and the commerce they support will struggle to operate.

- We will not be able to move or lift materials, dig foundations or drive piles; skyscrapers will stand monument to our ingenuity and our extravagant use of oil.

- We won't be able to make oil-based plastics.

The alternatives

Happily, pioneering folk have developed and used alternatives to oil for years. They have laid the foundations of an oil-free economy. We must understand what they have done and build on their achievements, quickly.

Introduction

What's it oil about?

NAME FIVE THINGS essential to you. Ask your family and friends to do the same. Did you say water, food, light, heat and love? Did you talk about monetary wealth, your house, car and other belongings? Perhaps you mentioned self-fulfilment, achieving your dreams, travelling and learning. Did you name oil? Everything you named depends on oil. Okay, oil can't buy you love, but it does bring you heat and light. You have oil to thank for your ability to travel and buy goods from any country, for putting food on your table and for the table itself.

From research labs to factories, from homes to communications and from transport to cheap food, oil fuels a way of life we find pleasurable and safe. We enjoy better health and broader experiences than our ancestors ever did. With notable skill and ingenuity, we transform oil. We make goods to wash our bodies and homes; decorative items; and the tools of trade and modern life. We have improved public health with clean water and medicine. We power communications that give us a social life from the comfort of our armchairs (or the back of the bus). We have entertainment on tap. We feed the human family using oil as fuel, fertiliser and pesticide. We have enjoyed economic growth.

We make consumer goods which are designed to have a useful life no longer than a few years or even a few hours. We burn fossil fuels recklessly, with little heed for maximising efficiency. We power the science and technology that give us new and increasingly inventive ways of using oil. We have altered the fragile balance of our planet, and now understand that we cannot burn over 80% of remaining fossil fuels without driving irredeemable change.[8]

The VFUU Price of Oil[9] challenges the assumption that we have to stick with oil to keep our modern, comfortable, safe way of life. It advocates a proactive decrease in the amount of oil we use; a move driven not by the faceless 'they', but by us.

In Part One, we see how oil seeps into everyday experiences and explore existing and upcoming alternatives to oil.

In Part Two, we ask 'Do we need to use less oil?'. We learn how oil takes with one hand even as it gives with the other, and reflect on the consequences of oil use on our daily lives, from the economic and social effects of unpredictable oil prices to local pollution and climate change.

In its final chapter, *The VFUU Price of Oil* proposes using our creative temperament and drive to reduce oil consumption while retaining the advances of the last 150 years.

Throughout, it celebrates our adventuring spirit, our quest to learn and develop, and our risk-taking and entrepreneurial skills. It shows how we can use these characteristics to shake off our oily habits.

This book is not an exhaustive thesis on any of the subjects covered and uses only readily available information. If you don't believe something, check. To make the book easier to read, the references for all data and facts are given in the Sources, which you can find online.

As we survey our relationship with oil, you will find things you never knew, technologies you didn't imagine, and histories and achievements that are rarely shown to be related. These interesting subjects are sometimes in footnotes and occasionally in sidebars, some have been pushed out the Appendices. You don't have to read them, but you will find out so much more if you do. Here's one now...

---

8 See "Where has all the oil gone?" if you can't wait to find out what that is all about.

9 Really, you don't want to see The Volatile, Fickle, Unpredictable, Undependable Price of Oil every time I refer to this book, do you?

How did we get here – was oil ever dependable?

From 1859 to 1939, we found many uses for oil and changed the lives of millions. Look at old photographs, watch movies, read newspapers and books. Fewer horses pulled carriages. Modern materials and modern marketing changed our homes and the way we dress. We started to travel farther in less time. The rich remained rich, and the poor gained access to materials that mimic the rare and fashionable.

Then World War II quickly absorbed all available oil. The war relied on oil to fuel transport, power factories and make materials. For example, all nylon production was diverted to the war effort, and US President Franklin D Roosevelt created the Petroleum War Council to ensure America could build and fuel its planes and ships. Throughout the world, the public experienced fuel rationing and the sudden absence of new, affordable materials.

With peace, public oil use resumed and grew, largely as a result of government policy. The Great Depression provided an enduring memory for British soldiers serving during the war. On their way to battle, the soldiers of the 50th Division asked Ernest Bevin, Minister of Labour:

> "Ernie, when we have done this job for you, are we going back to the dole?"

The wartime government identified economic success as the way to pay the moral debt owed to the men returning from war and the families of the fallen, so it acted to prevent economic slowdowns like those of the Twenties and Thirties. The resulting Employment Policy had 'a high and stable level of employment' as a goal, and employed several tactics:

➢ *Converting factories from the goods of war to the goods of trade*

➢ *Making production efficient*

➢ *Ensuring the workforce was capable, mobile and well paid*

➢ *Maintaining the balance of trade*

Governments around the world stimulated employment and economic growth. Oil powered these policies and fuelled a burst of consumerism. Furthermore, it allowed advances in many areas of our lives.[10] *The Fifties and Sixties were a golden era of prosperity. As the Seventies dawned, we consumed five times as much oil each day as at the end of hostilities in the Forties.*

Oil prices saw no such growth, and soon precipitated an astonishing series of events. When a massive devaluation of the dollar reduced the value of the money the oil-producing countries received, two organisations drew their members together to get higher prices in a market dominated by British, Dutch and American oil giants.[11] These organisations were the Organization of Arab Petroleum Exporting Countries (OAPEC) and the Organization of the Petroleum Exporting Countries (OPEC).[12]

In response to the 1973 Arab-Israeli war, OAPEC, along with Egypt, Syria and Tunisia (who were not yet OAPEC members) pushed up the price of their oil. Within days, they also stopped selling oil to the United States and all countries that supported Israel (for example the UK, the Netherlands, Portugal, Rhodesia-Zimbabwe and South Africa). They also cut production by 5% a month.

---

10 You will read the words 'after World War II' several times in this book. It has only been in this, relatively short, period that we have really learned to use oil to improve the lives and lifestyles of billions of people.

11 A brief history of the British, Dutch and American oil giants: Founded in 1909 and formally named British Petroleum in 1954, BP grew from a business based on the relationship between Britain and the Shah of Iran.

Royal Dutch and Shell merged in 1907 in an act of self-preservation against Rockefeller's Standard Oil to form Royal Dutch Shell, aka Shell.

Then the United States Department of Judgment sued and broke up the phenomenally powerful Standard Oil in 1909 under antitrust law. The resulting companies are familiar names today: ExxonMobil, Chevron, Sohio (now an arm of BP), Amoco, Conoco and ARCO.

A mêlée of these giants and smaller companies has seen mergers and takeovers, co-operation in one field and competition in others. By the Seventies, most major oil companies were British, Dutch or American, yet their home countries were responsible for only a small part of their production.

12 OAPEC consists of Algeria, Bahrain, Egypt, United Arab Emirates, Iraq, Kuwait, Libya, Qatar, Saudi Arabia and Syria.

Algeria, Angola, Ecuador, Iran, Iraq, Kuwait, Libya, Nigeria, Qatar, Saudi Arabia, the United Arab Emirates and Venezuela make up OPEC.

Within six months of the start of the embargo, oil prices almost quadrupled[13] and pushed up other prices. Workers in the UK's nationalised rail and coal industries held strikes to secure pay rises in line with rapidly increasing inflation. This industrial action combined with rising oil prices to create the 'energy crisis'. The UK government introduced the three-day working week to conserve coal stocks and went on to lose the 1974 general election. In this way, increasing oil prices caused hardship and led to a change of government.

Across the Atlantic, the American government fixed the price of oil from existing sources at a low price to encourage the oil companies to develop new sources. Instead, the oil companies withdrew old oil from sale, and new oil rose dramatically in price. We depended so much on oil that rising oil prices caused recessions worldwide.

When economies start to shrink, governments decrease interest rates. Back in the Seventies, instead of encouraging growth, low-interest rates worsened the economic effects of increasing oil prices. The period around the oil crisis may be the best-known stagflation event.[14]

A few years later, the Iranian revolution overthrew the Shah and installed Sheikh Khomeini as the Grand Ayatollah. Iranian production dropped by 7%, but Saudi Arabia increased its production and made up some of the shortfall. The oil price duly rose by 30%, though perhaps painful memories and the panic of motorists and the City were bigger influences than the 4% drop in global oil production. The impact across the world was stark: economic growth halted, and in the UK economic difficulties once again contributed to a change in government. This period is often referred to as the 1979 or second oil crisis.

---

13 There were lots of other political and economic reasons behind the embargo and rising oil prices, but this is a book about oil, so you will have to check them out yourself.

14 Stagflation occurs when prices inflate and the economy shrinks or is stagnant.

From its discovery until the early Seventies, we used more oil with each passing year. When prices quadrupled in 1973 consumption only dipped slightly. The economy and oil prices had periods of bad performance in the following three decades, and oil prices never returned to their steady state. They caused problems when they rose and problems when they fell. All the while, we continued our oily spending spree and doubled the amount we used from 1970 to 2013 (45 million barrels a day to 91 million).

Historic oil consumption and prices

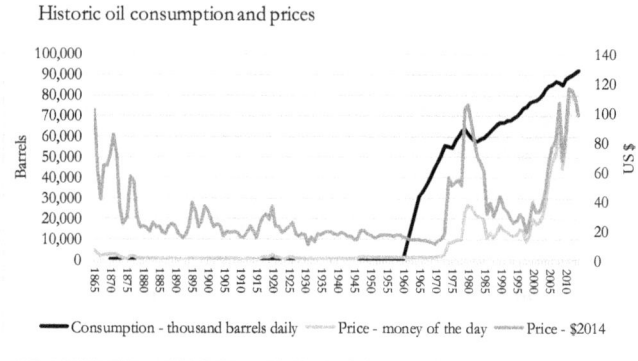

Figure 1: Historic oil consumption and prices

Few clouds lack a silver lining, and positive effects came from the realisation that the time of cheap oil had ended. In Europe and the United States, smaller, fuel-efficient cars came into favour. People developed and built alternative methods of producing electricity. In "Taking oil my energy", we will visit nuclear power stations, water barrages and other power plants built during the Seventies. Thanks to investments made then, and our experience, these technologies work well.

Just some quick notes on terminology. The many names given to oil, eg crude oil, petroleum, and condensate, make it harder to understand. So, other than when we need to use a technical term, 'oil' in this book refers any or all of the above. Also, it is far easier to type and read the single word 'oil' than the phrase *oil, the other fossil fuels, and their complex economic and technical relationships.* Unless the word 'oil' comes with a fact, you can take it to mean *oil, the other fossil fuels, and their complex economic and technical relationships.* For example, in "Plastic dolls" we talk about oil as the source material for plastics, but in the Americans mainly use natural gas. Again in "Plastic dolls" we say that 4% of oil is used to make plastics; in this case, as it comes with a fact, 'oil' means oil.

Part One

How do we use oil today?

IN PART ONE, WE WILL SEE that we are entirely dependent on oil, and in ways you might not expect. However, the best-understood use of oil is in transport so we will ease into the subject with "Here, there and everywhere". Transport needs manufacturing industry to make cars and roads, so in "Oiling the wheels of industry" we look at the ways manufacturing industry is reducing oil costs. Transport and many modern manufacturing techniques depend on plastics and we will learn about those versatile materials in "Plastic dolls".

Most of us understand that the price of oil affects the cost of energy. Because electricity is so darned useful, we will look at dozens of ways to produce it, without resorting to fossil fuels, in "Taking oil my energy". Then we will discuss a few topics closer to home in "Your goose is cooked", "Read oil about it" and "Homeward bound".

Of course, this part is only complete with a foray into the murky worlds of economics and politics, so we will then remember "It's the economy, stupid".

Each of these chapters gives a little bit of history, an overview of the way we use oil today and then a peek into the alternatives to oil. As we pass through these chapters, you will notice the same technologies mentioned in various circumstances. Each is an example of technology transfer. Transferring technology between applications – aka 'not reinventing the wheel' – is a common way of sharing learning and it benefits everyone. Keep your eyes peeled – indeed, come up with your own. My favourite technology transfer is from transport to energy production (using kites).

Chapter one

Here, there and everywhere

TRANSPORT IS EVERYONE'S favourite oil subject. After all, we almost touch oil when we fill our cars. Moreover, oil lets us see our friends and family. We use it to visit ancient wonders and encounter new cultures. We travel to enjoy more pleasant environments and take advantage of shopping and leisure facilities. Nevertheless, oil serves us best as the beating heart of commerce. We use oil to move goods from low- to high-wage economies, open new markets for products and services, and exchange raw materials and skills. In this chapter, we will look at our primary modes of transport and ask how they can be less oily. You will find third-generation biofuels, hydrogen-powered cars and sailing ships – all of which have a part to play in the future of transport. But first, let's run through the history of transportation, how it uses oil and its economic importance.

The history of transport, how it uses oil and its economic importance

Trade is the foundation of our economy. It built empires, drove migrations, sparked wars and demanded international law. Whether it was the Egyptians importing spices from East Africa and Arabia, or the Romans using sea routes to reduce the cost of moving goods around their empire, transport drove trade – and that spread human development. Whether it was the Western nations vying for wealth in the East, or the Spanish looking for a shortcut to the Indies, transport drove commerce – and that shaped our culture.

Closer to home, trade and commerce led to our grandest public projects. The need to move goods to distant markets advanced the canal network, which rail then rapidly overtook as the transport of choice for coal, grain, stone and passengers. Without high-volume freight transport, the Industrial Revolution would have stopped in its tracks. Then the internal combustion engine came along and delivered our ability to choose the road we follow.

Transport now uses oil in a myriad of ways. Fuel is the most obvious. Petrol, diesel and kerosene are the oil industry's biggest earners and they consume around three-quarters of each barrel of oil. Petrol accounts for about half of every barrel, diesel about a quarter, and kerosene one-tenth.

Moving on, concrete and iron are thirsty for oil, and roads, rails and docks use vast amounts of both. Heavier hydrocarbons, with coal and gas, power the immense factories dedicated to making vehicles and marine vessels. Lighter distillates become plastics, making vehicles lighter and safer. The all-important fashion features that sell transport and give brand status to its makers and providers rely on the versatility of hydrocarbon chains.[15] Navigating from A to B depends on oil, as we make satellite navigation units from oil and they connect to satellites that are oil-expensive to design, build and launch. We will think about the use of oil in manufacturing in "Oiling the wheels of industry".

When you last took home a new purchase or had it delivered, you probably tore, cut and ripped through layers of plastic and card. All that packaging provides a secure shell for goods that travel. It cushions drops, supports the weight of stacked boxes, protects against humidity and extreme temperatures, and thwarts thieves. If you fancy some salad or other fresh food, a plastic bag protects the product, gives information and shows off the brand. Take the shrink-wrap on cucumbers. The

---

15 Oh my word, so much jargon so early. These terms are explained in "Plastic dolls" and in Part Two. But in case you can't wait: oil is a mix of molecules made from hydrogen and carbon (in chains); some chains are short and light, others are long and heavy. Fuels such as kerosene (mainly used to fuel aeroplanes) are made up of medium-length, medium-weight chains. The chains are separated in a process called fractional distillation.

plastics industry tells us the plastic sleeve increases the shelf life from three days to ten days, saving energy and preventing waste.

Consumer packaging not only protects goods, but it also sells the product it contains. If you buy CDs or DVDs, note how publishers use the case to show off the music or movie. If you indulge in high-end cosmetics and toiletries, the plastic pot weighs more than the product and the beautifully presented box looks like gift wrapping.

And of course, packaging uses oil in manufacturing and as a raw material. Between over-packing and consumer packaging we use 39% of the plastic made in Europe to protect products. We will think about plastics in "Plastic dolls" and learn where all that packaging ends up in "Where has all the oil gone?".

Just as transport depends on oil, we depend on transport. In the UK, transport and storage businesses employ 5% of the working population, with the biggest concentrations around Gatwick and Heathrow. Then we need to add in the people making vehicles and building roads. Millions more depend on transport to operate their business, services and hospitals. We can't stop moving goods, and we don't want to stop travelling, so what can we do? Before we move on to the all-important question of how to fuel transport without oil, let's look at the particular needs of freight transportation and the ways the industry is already reducing its oil consumption.

Saving a packet

Speed and efficiency are essential for cheap mass goods transport. Our need for both drove the development of the industrial container – those large steel boxes you see at every dock, on trains, and behind warehouses and factories. The first modern containers carried goods in 1956, but container-type boxes have been around since the 18th century.

Containers improve speed and efficiency (ie reduce costs) in several ways:

- They hold thousands of boxes or several pallets, which would otherwise need separate handling every time they change mode of transport.
- They come in standard sizes, allowing manufacturers economies of scale for handling equipment, trailers and ships.
- They protect goods from damage and theft.

- They stack and pack with little wasted space.
- They provide an enclosure for goods, so reducing lorry and train costs.

Understandably they are very popular. Indeed, the world owned the equivalent of 32.9 million operational twenty-foot containers in 2012 and built a further 1.6 million the following year. Over 6,000 container ships carry more than 20 million steel boxes and the largest container vessels in service hold 18,000 containers though the width and depth of waterways constrain further growth.

You might ask why *we* should care, well… weighing 2,250kg each, the oil cost of making the steel for each container is more than 19 barrels of oil or just over 3,000 litres. And all that weight burns even more fuel.

Containers are very visible, but few of us see the oil used to keep stuff cool while in transit and storage. Transport experts call the activities, facilities, equipment and vehicles used to carry temperature-sensitive items the 'cold chain'. The cold chain gives a cold environment from the moment the pea pod goes pop to getting a bag into your trolley at the supermarket. It controls the temperature in transit and storage and enhances the shelf life of goods such as chemicals, drugs and radiography supplies. To maintain the right temperature, lorries, trailers, containers and warehouses have cooling units, insulation and electronic monitors to track goods during every step of their journey. For example, Radio Frequency Identification (RFI) tags show the location of goods and tell logistics firms and their customers how long their cargo is spending at the right or wrong temperature.

Companies use central warehouses to maximise profits. They unload containers, store goods and dispatch orders. These warehouses minimise businesses' inventory costs and provide a rapid response to demand. On the downside, they increase transport miles. A report commissioned by DEFRA cites centralised warehousing as a driver in the rise of food miles though they also state that using fuel-efficient HGVs (instead of small vans) may offset the environmental impact. In "Your goose is cooked" we will see that food miles are a crude measure of sustainability; however, extra miles do increase the oil costs of transport and the cold chain.

Concern for the environment and the high oil costs of the last couple of decades have driven manufacturing and transport businesses to reduce packaging costs, through clever design or by reusing packaging when possible. For example, one food producer found changing the shape of cereal boxes did not lessen the volume of the box, but did reduce packaging by 36%. In making the change, they saved tons of card and thousands of barrels of oil. Other manufacturers reuse packaging. You have probably seen bread crates outside supermarkets and factories reuse plastic boxes to move parts, a practice that started in earnest with automotive manufacturers.

Some businesses are adopting non-oil packaging. The bioplastics we will meet in "Plastic dolls" make expanded foam packaging and peanuts. A leading computer manufacturer uses bamboo and mushrooms to make boxes and fittings for laptops and is experimenting with wheat straw. You will also find bioplastics protecting magazines and greetings cards.

The way we choose to move goods has a significant effect on the amount of oil we use, particularly when you take into consideration the amount each mode of transport moves. For every tonne of goods[16] moved one kilometre, the carbon dioxide (CO_2) emissions[17] are:

- Air (domestic) 2.85kg CO_2
- Air (short-haul international)1.21kg CO_2
- Air (long-haul international) 0.67kg CO_2[18]
- Road (average diesel van) 0.53kg CO_2
- Road (average laden refrigerated HGV) 0.13kg CO_2
- Rail (freight train) 0.03kg CO_2
- Sea (average general cargo ship) 0.01kg CO_2

Taking freight out of the air and off the road are the quickest ways to reduce oil costs, but any business that changes transport mode will experience changes in its cash flow. Difficulties occur when outgoings

---

16 You can't really equate this to the costs of moving people around, as we take up more space per kilogram (unless we are on a commuter train).

17 Getting hold of actual consumption numbers is difficult and they rarely add up, so throughout *The VFUU Price of Oil* we make an assumption that lets us use data collected and collated for climate science, which is *carbon emissions = energy = oil*. We use the assumption many times and point out when it is invalid. In this case, the assumption is pretty good.

18 The effect on the environment of flight is about double these numbers because of something called radiative forcing. But radiative forcing makes the assumption in the previous footnote false, so it is not included.

increase (for example, to buy new equipment or pay contracting costs) before savings start. The European Commission used to run a programme called Marco Polo to give financial support during a 'modal shift' (eg taking freight off the road). The programme gave grants to 'commercial undertakings' (companies) for reducing or avoiding road transport, adopting best practice and innovating freight transport. The Commission released new funding annually. The programme ran from 2006 to 2013.

Over ten years Marco Polo cost 552 million Euro. The programme supported 650 companies and they removed 41.4 billion tonne-kilometre of freight from the roads (these benefits will increase because the project's final report doesn't cover the outcome of incomplete work). Companies can now find similar support through the Connecting Europe Facility. Replacing Marco Polo saved funds and created a broader scope though the loss of the original name seems a shame.

Fuelling the future

Personal transport and freight have fuel as a common denominator. On the road, individuals travelling by bus, motorbike, taxi and car burn 63% of our road transport fuels. Heavy and light goods vehicles (lorries and vans) use the remaining 37%. So let's look at powering personal transportation, remembering everything that applies to ferrying us about also applies to freight.

Automobiles (including buses and motorbikes)

Substituting petrol and diesel with an identical fuel would be an easy transition. We could keep our cars and behaviours, right down to filling up at the 'petrol' station. What we lack are fuels we can exchange for oil-based petrol and diesel. Most current alternatives demand changes to our cars and a new supply network. Fuels that come close to the holy grail of transport power are other fossil fuels (LPG and natural gas), biofuels and hydrogen.[19]

Let's run through each of those alternative fuels. Liquid Petroleum Gas, or LPG, is the common name for propane and butane. We can buy the gases separately or as a blend. According to legend, a chemist called Walter Snelling first identified propane. His discovery came when a

---

19 That said, advances in electric vehicles mean they do now offer a similar experience.

motorist complained half his gasoline evaporated on his journey from the gas station to home. The evaporating fuel was propane, butane and other gases.[20]

The primary costs of turning to LPG are converting cars (or buying new) and building filling stations. These costs affect personal drivers and commercial drivers in different ways. For example, the owners of large fleets can build gas stations while private drivers need to find them. With this in mind, you can understand why the United States' move to LPG has started with fleets of lorries owned by companies and governments.[21] Outside the United States, LPG is reasonably common and drivers have good support for conversions and fuelling. America has nearly 3,000 public and private LPG fuelling stations, the UK has more than 1,400, and there are 27,000 across the rest of Europe.

We also use natural gas (methane) to fuel cars, and, for once, renewable and petrochemical industries are on the same side – sort of. Both actively promote natural gas, and, as a result, government departments, companies and private drivers are adopting Compressed Natural Gas (CNG) and Liquid Natural Gas (LNG). The oil industry gets its methane from natural gas and oil wells, whereas the makers of renewable natural gas use landfill gas, wasted food, farm waste and sewage. Again the US is following Europe in this trend. Sweden, Switzerland and Germany all have impressive CNG fleets. These countries mainly use biogas (largely from pig manure) to power their cars. The UK, too, lags the continent, but British companies have started to fuel lorries with methane, and biomethane supplies are growing.

Fossil LPG and non-renewable natural gas are finite and their costs will rise. However, moving to these fuels reduces both dependency on crude oil and carbon emissions. Therefore, they buy us time to develop fresh ways of motoring on affordably.

In the early days of the motor industry, liquid biofuels powered cars alongside oil-based fuels. Back in the day, oil won for several reasons:

- Petrol has a higher energy density than primitive biofuels.

---

20 With that nugget of information, you may have realised that the oil and natural gas industries produce LPG as a by-product. The growing popularity of LPG also justifies its own wells.

21 Try to bear this in mind when you read "To oil, or not to oil?".

⊺ Petrol production is not seasonal, nor do you have to worry about bugs or losing crops to drought.

⊺ Petrochemicals use less land.

⊺ The materials employed by the early motor industry worked better with petrol than biofuels such as ethanol.

⊺ Oil companies had a strong reason to make petrol the fuel of choice: there was no competing biofuel industry.

Now we recognise the advantages of biofuels. Second-generation biofuels match or beat petrol on energy density, and they come from non-food crops and waste, which replenish in months rather than millions of years. To compete with oil-based fuels, companies are coming together in an evolving industry.

Alcohol was an early biofuel. The Model T ran on it and Henry Ford believed:

> *"The fuel of the future is going to come from fruit like that sumach out by the road, or from apples, weeds, sawdust – almost anything. There is fuel in every bit of vegetable matter that can be fermented. There's enough alcohol in one year's yield of an acre of potatoes to drive the machinery necessary to cultivate the fields for a hundred years."*

Today – Ford's future – many cars use alcohol because petrol contains ethanol. Ethanol levels are expressed as an 'E' number; eg Europe's is E5, meaning we have 5% ethanol and 95% petrol.[22] Typically the US adds 10% (E10). The Australian government limits ethanol content to 10% by law, while Asian motorists enjoy a broad range of blends. Some cars can run on higher blends; for example, 85% ethanol (E85) and pure ethanol (E100) are becoming more popular.

Brazil, the most advanced sustainably-fuelled country, uses pure alcohol and blends. Its fuel contains at least 18% alcohol (E18) and Brazilians can no longer buy pure petrol. In other countries, environmental laws drive the sale of ethanol in petrol. For example, in the EU 5% of all fuel must come from renewable sources by order of the Renewable Energy Directive. Most UK fuel producers blend diesel with biodiesel to meet the target. In 2020 the percentage of renewable

---

22 Similarly, 'B' indicates biodiesel and 'R' stands for renewable diesel.

fuel mandated in Europe will increase to 10%. Petrol companies are likely to increase ethanol content in the lead-up to the target date.[23]

Adding ethanol to fuel has significant drawbacks. First, ethanol damages older cars. Second, ethanol and petrol burn differently so makers and drivers cannot tune cars to give their best performance. Third, today's ethanol is a first-generation biofuel and comes with high oil and environmental costs. Outside Brazil, the typical feedstock for alcohol is maize, and producing it uses more energy than it releases though there are claims and counter-claims about these numbers. And maize is a monoculture, which creates issues in areas where farming is concentrated.

However, adding bioethanol to fuel today could be a case of one step back to take two steps forward, as working through the drawbacks will make the supply chain and cars are ready when second-generation fuels come on to the market. Still, if the US government had nourished a silver buckshot[24] approach, we may already have those second-generation fuels.

We should mention in passing that some cars run on methanol, a form of alcohol with one carbon atom. Nowadays these are mainly high-performance vehicles, like drag racers. If George Bush Senior had had his way in the late Eighties, methanol would be more common. He proposed banning petrol-burning cars in city centres to remove local pollution. Of all the alternatives we are considering, methanol got the President's vote. However, methanol releases less energy than ethanol and causes more damage; disadvantages which stop it competing outside specialist fields. Also, at the time, the best source of methanol was coal…

We mentioned that the petrol companies are meeting biofuel targets by adding biodiesel to diesel. In "Plastic dolls" we will come across a chemical process called transesterification. It is jolly useful. When used on waste cooking oil, it swaps an ester molecule with alcohol, and biodiesel pops out. Making fuel this way is easy; enthusiasts call it 'home brewing'. Biodiesel (a first-generation fuel) has a lower energy density than diesel, and it may damage engines designed with only diesel in mind.

---

23 The US takes a different route: The Renewable Fuel Standard sets a volume of ethanol that companies must sell each year.

24 This is a teaser – 'silver buckshot' will be explained in "Taking oil my energy".

But the small volume of biodiesel added to European diesel does not damage cars, as it is a second-generation biofuel. (Don't worry – we'll get to that soon.) Biodiesel may produce more NOx[25] and smaller particulates than oil-based diesel.

However, (and you were expecting it) most of today's biofuel comes from plants we could eat or crops grown in place of food. If we stick to growing biofuels, they cannot be a large-scale alternative to fossil fuels. Brazil is the exception; it makes ethanol from sugar cane, and it has vast areas of open land and enjoys much sunshine.[26] Clashes between food and energy are growing more frequent and have contributed to the food crises we will discuss in "What's the worst that could happen?".

But we shouldn't write off biofuels. Our friends the scientists are not sitting still; they are developing and starting to produce second-generation biofuels. These modern fuels need less energy, use less land, and interfere less with food crops and the ecosystem than traditional biofuels. To make biofuels as good as, or better than, fossil fuels, developers borrow manufacturing techniques from the oil industry – but there the parallels end. Second-generation fuels use non-food plant crops and municipal waste as feedstock. The work as well as or better than fuels made from oil.

Second-generation biodiesel, or green diesel, has a higher energy density than fossil fuel diesel. It flows better when cold, and has roughly the same sulphur content and a higher cetane rating, which means it lights more readily. Despite improved characteristics, burning green diesel in older cars poses no problems, because it's similar to the old stuff. You might hear it called renewable diesel. However, oil and gas companies add hydrogen to bio-feedstock and blend it with fossil fuel diesel in refineries and that hydrogen comes from natural gas, so 'renewable diesel' will not live up to its name until we use renewable

---

25 Yup, you say it how it looks. It describes the various gases made when nitrogen and oxygen combine – eg NO_3 or N_2O – each of which has a specific (and not good) effect on the atmosphere. See "Where has all the oil gone?" for more.

26 However, even Brazil is seeing a disruptive effect from combining the food and fuel markets. Rising sugar prices mean farmers are keen to sell their crops as food instead of fuel.

energy to produce that hydrogen from water.[27,28] What's more, as we make it the same way, we can assume it is as bad as oil-based diesel in producing NOx and particulates.

Another method of making second-generation diesel starts with gas. As well as the other benefits we will see; gas-to-liquid diesel is less polluting than oil-based, first- or second-generation biodiesel. To make synthetic gas ('syngas') we capture carbon dioxide – from, for example, a power station or steel plant – then convert it to methane. We have umpteen methods of creating syngas, and new ones in development. Improving how we make syngas reduces production time and energy costs, and gives better products.

You would be right in thinking a common source of carbon dioxide is rotting plant matter. Unfortunately, this gas is not easy to capture as that from industry, so, to use plants as a source for syngas, we need more science. Getting energy from woody stems is difficult – which is why we do not eat them – and first-generation biofuels used foodstuffs.[29] To take full advantage of plants we need the energy in their fibres. Gasification serves this purpose.

The method is well established; gasification made town gas in the 19th century and is the first step in making liquid fuels from coal. It works by heating the fibres, made of cellulose, which splits them into light hydrocarbon gases (syngas). Air or steam sometimes aids the process. Pyrolysis is the name for a kind of gasification without oxygen, and a similar process called torrefaction operates at lower temperatures.

Now we have the syngas; the Fischer-Tropsch Process turns it into green diesel or other fuels, (eg methanol) or chemicals (eg dimethyl ether).[30] (We will meet Herren Fischer and Tropsch and gasification in more detail in "What has the oil industry ever done for us?".) These fuels build on proven techniques, so they are no more complicated than

---

27 Or, slightly ironically, methane.

28 In "What has the oil industry ever done for us?" we will learn about hydrogenation, the process that removes sulphur from oil by adding hydrogen. So-called renewable diesel is made using the same method.

29 It is also the reason why animals that eat tough vegetation have complicated digestive systems, graze continually, and are not good at running quickly over long distances.

30 Of course we could just use the gas, as the Bio-bus in Bristol does. After treatment, the 'waste' from five people can provide the fuel for 300km of passenger transport.

extracting oil and producing petrol, diesel or kerosene. Sweden and Italy have commercial second-generation plants; the US has a round dozen. Ironically, Karl Diesel invented the diesel engine with biofuels in mind; like Henry Ford, he must be wondering what we have been doing for the last 80 years.

Though, alcohol has also progressed to a second generation. If you want to make alcohol from plants (to power cars), you split cellulose into sugar. Do you remember an experiment at school in which you spat into a test tube containing starch and iodine? The starch turned the iodine purple, but an enzyme in your saliva converted the starch to sugar, and the colour faded away. In 2004 Canadian firm Iogen turned this classroom demonstration into a working factory (with cellulose instead of starch), which used enzymes to convert farm waste into sugar on an industrial scale. Making and distilling alcohol from sugar is the same as making it from grape juice or malted barley.

Iogen's Brazilian partner, Raizen, has built a plant to make alcohol made from the parts of sugar cane that otherwise go to waste. The new facility can produce 40 million litres of ethanol per year with no extra land, fertiliser or pesticide use, and they have four more projects in the pipeline.

Oil splits opinions, but it is not as divisive as genetic engineering. In 2010, when Craig Venter announced his labs had created synthetic life, many sensational news reports missed his primary goals. He created simplified bacteria, which will eventually make chemicals (oil?), decompose pollutants (hydrocarbons and plastics?) or make proteins for vaccines. But we don't have to look to scientists as controversial as Venter to see microbes making fuel. Others have shown we can make oil from carbon dioxide without creating life though sometimes they use genetic engineering. They also use more familiar chemical engineering techniques.

In 2010, the United States Department of Energy's Joint BioEnergy Institute (JBEI) announced it had created a strain of E. coli that makes diesel from carbon-rich waste. E. coli is the humble bacteria that lives in all our guts and occasionally causes bouts of severe food poisoning. More recently, in 2013, Korean scientists declared they had made petrol using modified E. coli. At the University of Minnesota, researchers have made fuel from carbon dioxide using cyanobacteria.[31] (Synechococcus) and bacteria (Shewanella). Moving away from bacteria, a fungus (Gliocladium roseum) lives in the jungles of Patagonia and breaks down cellulose, making hydrocarbons similar to those found in diesel. These discoveries have shown we can engineer fuels using microorganisms, and some are production ready.

Lanzatech, an American/ New Zealand company, will soon power Virgin Atlantic planes. They make their fuel by fermenting waste gases from steel production.[32] In Australia, Algae Tec's systems will use algae to convert carbon dioxide from power plants and factories into feedstock for biodiesel and animal feed. Working with the producers of carbon dioxide gives both sides an advantage: Algae Tec get one of their main ingredients and the producer meets its emissions targets. The latter achieves direct financial benefits in countries with a carbon tax.

Equally, going 'old school' could be part of our oil-free future. In the Seventies, biofuels were more straightforward; scientists converted whole algae into hydrocarbons. Then the tide changed to extracting fats from algae, but extracting the energy-rich fats consumes a lot of energy. Now scientists in the US Department of Energy's Pacific Northwest National Laboratory have returned to earlier methods. They have shown we can crush algae to make feedstock for petrol, diesel and aerospace-grade kerosene. In "Taking oil my energy", we will talk about algae towers. The algae housed in these towers could make biofuels or hydrogen, with less use of land than biofuel crops and reduced transport to consumers.

Specialists call biofuels made from algae 'third-generation'. The name will probably cover bacteria and fungi (passing quickly over the joke about the mushroom who goes to a party) – though these might be fourth-generation. Let's stop before we need to draw a family tree.

In contrast to the complex chemical make-up of biofuels, hydrogen is the simplest of substances and is a good energy store, so let's get stuck in. Hydrogen (H_2) molecules consist of two atoms and exist as a gas at

---

31 Cyanobacteria are bacteria that use photosynthesis to produce energy. They live almost everywhere and undertake 40% of all photosynthesis. Without them we would not have an oxygen-rich environment. Although they are better known as blue-green algae, do not confuse them with algae, which are plants.

32 They are keeping the nature of their microorganisms secret.

room temperature and pressure – until they combine with other elements, which they quickly do. Indeed, H_2 is too reactive to exist naturally on Earth. But when we separate hydrogen from carbon (hydrocarbons) or oxygen (water) we create an energy store. Three technologies use hydrogen: fuel cells, internal combustion engines, and rockets.

Fuel cells have been around since 1838 and 1839 when two physicists developed them independently in Germany and Wales. They came of age when they powered the Apollo flights (fuel cells, not the physicists). A couple of decades later, the company that made the fuel cells (UTC) started to make large, stationary fuel cells to heat buildings. Around the same time, in 1991, Roger Billings built the first car powered by hydrogen fuel cells.

Some fuel cells use hydrocarbons instead of hydrogen, but we will skip over them for obvious reasons. Fuel cell technology is complicated and several types of cell are in development or use. However, they share three basic steps:

1. The hydrogen is split into a proton and electron (nothing like splitting an atom; it uses simple chemical reactions).
2. The electron whizzes round a circuit to the other side of the fuel cell, creating a current. The proton takes a more leisurely route through a material called an electrolyte.
3. When the electron and proton meet, they recombine and react with oxygen.

Roger Billings is a hydrogen energy pioneer. He experimented with hydrogen internal combustion engines in cars for nearly 30 years before he used a fuel cell. He followed a long tradition. François Isaac de Rivaz built the first hydrogen-powered car in 1807 though commercial interest in hydrogen ran dry until the late 20th century and its high oil prices.

Since François de Rivaz's first hydrogen-powered car, manufacturers have produced more than 200 hydrogen-powered vehicles, most since the Nineties (some are not cars). Historically, hydrogen-powered cars have all been prototypes or part of short production runs. But the world is changing. Toyota launched its Mirai, powered by fuel cells, in Japan in December 2014; Hyundai started sales of a fuel-cell version of its ix35 in 2015, and Mazda has been making a hydrogen hybrid since 2009. They are not alone; many major car makers have plans to introduce hydrogen-powered vehicles. Of course, the infrastructure to support them is also required – currently you can fill up at 184 fuelling stations worldwide[33] (the UK has 17). Norway has a hydrogen network, which was initiated by Statoil, and has joined with Sweden and Denmark to support the introduction of clean technologies.

While we are talking about hydrogen, let's think about rocket fuel. You might believe it is not relevant to a chapter about mass transport, but shortly you will see rocket power could be a valid alternative to oil-based flight. In rockets, liquid hydrogen and oxygen mix, then ignite. As the gases combine to form water, they produce heat and expand. The expansion creates a jet and, following Newton's third law of motion, the escaping jet exerts a force on the rocket, which causes it to move forward. (Beware erroneous claims that the gases push against the ground.)

The disadvantages of hydrogen are similar to biofuels and petroleum gases: The infrastructure is still developing, our cars will not burn H_2 and we need energy to make it. We can overcome these issues. We have seen that an infrastructure is growing and commercial vehicles are available. Likewise, the renewable energy sources in "Taking oil my energy" suit hydrogen production.

Let's talk about the elephant in the room. We have grown up knowing hydrogen is explosive and transporting hydrogen is risky. Yet industry uses millions of tonnes.[34] The supply chain is robust and safe, and new uses of hydrogen will build on the knowledge and best practice of today's producers and users.

Biofuels and hydrogen weren't petrol's only early competitors. Until the Thirties, petrol-fuelled vehicles had several problems – including the smell, vibration, difficulty in changing gear and the crank start. Only when manufacturers resolved these issues did petrol pull away from electrically powered cars and vans. Now electric cars suffer in

---

33 Operators have plans to build a further 129, a third of which will supply 'green' hydrogen, ie that made from water using renewable energy.

34 The list of uses is long: The biggest use of hydrogen is ammonia production, and as one of the biggest uses of ammonia is creating fertilisers, hydrogen feeds us. H_2 enables oil refining processes and their biofuel equivalents. It protects food, welds metals, makes lasers, detects leaks, cools power stations and fuels nuclear fusion. It is an ingredient in chemicals, eg hydrogen peroxide, hydrogenated fats, lubricants, cleaners, cosmetics and vitamins. It makes electronics, glass and methanol and extracts metals from their ores. Phew.

comparison with petrol and diesel cars because they cost more to buy and the charging infrastructure is immature. However, the United States boasts almost nine times the number of stations and charging outlets than LPG and CNG stations combined, and they are springing up all over the UK.[35]

All electric cars can cover the UK's most common journeys[36], most for several between charges, and many homes and businesses can charge them. We can turn to electric vehicles (EVs) in our thousands. Looking at this cold logic might make you think EVs are on their way, but convincing the British driving public to go electric remains an uphill challenge. France, however, enjoys a considerable advantage over the gas- and coal-powered UK. They produce most of their electricity in nuclear power stations, making travel in an electric car environmentally attractive (and cheaper than in the UK). Low carbon emissions, coupled with zero point-of-use pollution, make electric loan cars in Paris a no-brainer. 15,000 people subscribed to an EV loan scheme in the first six months back in 2011. Their subscriptions give them access to Bluecars, which they pick up and return at any rental station in Paris. As well as benefiting Parisians, the scheme proves technologies and a business model, and promotes the principles of car sharing. It is also pretty egalitarian as private owners of electric vehicles can charge at the stations – for a small fee. Bolloré, a French industrial and investment group, manage the scheme and call it Autolib'. After its success in Paris, it moved on to Bordeaux, entered the American market via Indianapolis 2015 and took over an existing electric vehicle system in London in 2014.

The Bolloré ventures are a bold drive into the vexed subject of electric cars. In the UK we have a bit less action, but plenty of evidence that this is a real alternative to oil-powered transport. As early as 2010 a project called CABLED brought together car makers, three universities, two councils and an electricity supplier (and a partridge in a pear tree) to understand how to improve electric vehicles and understand driver behaviours. More recently, Highways England has assessed the feasibility of installing Dynamic Wireless Power Transfer systems on the strategic

---

35 You might wonder about solar panels, with good reason. Although solar panels power concept cars and supply electricity in some mainstream vehicles, they cannot drive a road car – yet.

36 78% of England's car and van journeys are less than ten miles.

road network. So much interest and research can only result in better cars and increased take-up.

When we listed fuels, back in the introduction to this section, we didn't mention efficiency. But we will see efficiency throughout *The VFUU Price of Oil*, as using less oil is often the cheapest way to reduce our consumption. In road transport efficiency comes in three ways; the first (and most obvious) is better fuel consumption in vehicles. While we already know that personal road transport uses more fuel, changing cars runs into the problem we explore in the second part of the appendix "The car dilemma" – namely, unless you do the mileage, you could use more oil in building a new car than you save driving. However, if you run a large fleet of trucks, you are certainly clocking up the miles and are likely to change vehicles regularly, making adopting fuel-efficient technologies a 'no-brainer'.

At least, that's the thinking behind the combined efforts of two US government agencies: The Environmental Protection Agency (EPA) and the National Highway Traffic Safety Administration (NHTSA). These groups are responsible for a two-phased approach that will see a doubling of fuel efficiency in trucks built between 2014 and 2018. An enormous range of changes – from adopting the alternative 'fuels' we have discussed, through to improved aerodynamics and tyres – will yield oil savings equivalent to a year's (US) residential energy, or a year's (US) oil imports from OPEC.

Returning to cars, the second equivalent of energy efficiency is not driving. In most American states, driving licences expire after four years, and drivers must reapply. But young drivers are not doing so. The number of driving licence holders is decreasing and the number of miles driven annually has dropped for seven consecutive years. Reductions in car use may be due to youth unemployment and increasing virtual contact, and new 'transit' systems may have taken drivers off the road, as they help people forgo the expense of running a car. Patterns of car use see similar changes in other countries; for example, in the UK, costs are the main reason young people do not have driving licences, and the same follows in Japan. At the same time, as people in industrially developing countries climb out of poverty, they aspire to exchange walking for cycling. Then they upgrade pedal power to petrol with motorbikes and cars.[37] So supporting public transport at home and developing it in other countries may be the only way of avoiding this growing demand for oil.

And that third form of efficiency? Eliminate the driver. You don't need to be a mathematical whizz kid to know that half of us have worse-than-average driving skills, even though we all think we're the best. Most of us drive in ways that burn excess fuel. If the companies investing in self-driving cars get their way, the human factor in oil consumption will disappear. You might not like the idea. You might think it will never happen. Still:

- Google cars have clocked up thousands of independent miles.
- Nissan plans to have driverless cars in Japanese showrooms in 2020.
- Australian autonomous trucks are set to escape from mines and hit the highways.
- A Dutch shuttle bus has become the first driverless vehicle to be allowed on public roads – in the world.
- Driverless shuttle pods (based on those we shall shortly see at Heathrow) will be trialled in Greenwich in summer 2016.
- The UK government has put aside £20m to research driverless vehicles on 40 miles of British road.

When kicking the oil habit, what goes for cars goes for buses. Buses with fuel cells and hydrogen-powered internal combustion engines are in use and on order in many countries. In Milton Keynes (and Korea), electric buses are in service; their batteries charge when they drive over special plates in the road (the same idea as the Dynamic Wireless Power Transfer systems assessed by Highways England). As these technologies are familiar, let's spend some time on the rails to understand how public transport can get us out of our cars.

Trains

Ironically, our desire to take to the air often drives advances in rail transport. In this section, we will visit two examples. When we fly, we often rely on transport provided by the airport or a third party to reach

---

37 In industrialised countries the opposite is true – one might even say two wheels good, four wheels bad. Cycling and walking probably deserve their own section, but you know all about them…

the terminal. Once in the airport, we face enormous, sprawling buildings, and time is often scarce. Airport transport takes high volumes of people from A to B efficiently.

Magnetic levitation (maglev) earned its pedigree at Birmingham International Airport in 1984. Magnets lift and drive the train. The repelling forces between magnets in the tracks and train raise the carriages. Further magnets propel the train, by attracting it from the front and repelling it from behind. The magnets change polarity as the train passes over them. The United States, Japan and China all have maglev test tracks. Passengers travel on maglev in Aichi, Japan (the Linimo); Shanghai; China; and at Incheon Airport in South Korea. However, the lack of economic growth in recent years has made the high capital costs of many maglev projects less affordable.

While some maglevs are monorails, don't confuse the two. Most maglevs have twin rails, and monorail cars tend to connect with their tracks. Monorails are an older concept; the first greeted passengers in 1825. The world has 53 working multi-station monorails, with dozens under construction or proposed. Unfortunately, monorails have a disadvantage. It's difficult to switch trains between rails, meaning they're best for round journeys or trips along a single line. On the upside, monorails use energy from any source, reduce car traffic, and are ideal for public spaces, such as amusement parks.

Personal Rapid Transit (PRT) does not always run on rails but does need guides. The Heathrow PRT runs between a pair of kerbs. The United Arab Emirates, United States and South Korea also have PRTs. Pods take passengers direct to their destination. They might follow a single route, as does the PRT built for the 2013 International Garden Festival in the Republic of Korea. Others serve multiple pickup and drop-off points, such as the PRT network in place and growing in Masdar City, Abu Dhabi.

PRT pods carry no more than six passengers and sometimes that isn't enough. Group Rapid Transit (GRT) uses larger pods, which look like buses or minibuses. The Dutch city of Capelle aan den Ijssel runs a GRT between a subway station and a business park. Nearer to home, Daventry Council is planning a GRT and Cambridgeshire has a very successful scheme called the Cambridgeshire Guided Busway, which opened in 2011.

Rapid transport systems are one of the railed systems that are lighter than traditional trains. The others include automated trains, light railways and trams. As you might guess, these are collectively known as light railways. They are more prevalent in urban areas, with some outreach to suburbs and, occasionally, rural locations. Unlike heavier railways, which are under pressures of capital costs, light railways are undergoing a resurgence that is reducing car use in cities.

In the 30 years after World War II, the only countries to build light railways that still run regularly are:

- Russia (which has 42)
- The (at the time) Soviet republics of Belarus (one), Latvia (one), Kazakhstan (four) and Ukraine (13)
- The (at the time) Warsaw Pact countries of Poland (two) and Romania (two)
- China (one)
- Croatia (one)
- The United States (two)
- Germany (one)
- Japan (eight)

From 1975, tramways grew more popular in Western Europe, and the turn of the century saw growth on other continents. Americans tend to prefer heavy rail, eg metros and subways, so the US has relatively few light systems.

When it comes to public transport, technology is willing but the flesh is weak. Reducing our dependence on the motor car is easy, yet in England in 2014 we made about two-thirds of our journeys in cars as a driver or passenger. To make public transport attractive, governments employ carrot-and-stick approaches. Councils find charging for car parking generates revenue and pushes some motorists on to buses – the stick. They run Park and Ride schemes to provide the carrot. These systems are designed to keep visitors coming by providing them with a safe, secure, cheap alternative to driving into town and city centres. However, the designers of public transport sometimes seem to forget the traveller. For example, by focusing on reducing air travel, the introduction of high-speed rail has increased fares for rail passengers in Europe. Still, if they build it[38] (undependable oil prices mean), we will come.

Planes

From the moment the Wright brothers put an engine in their glider, the aptly named 'Flyer', aviation has fuelled the imagination of engineers and explorers, and given a lift to business. Without flight we would eat a smaller variety of fresh food. We couldn't take as many foreign holidays. We would have to make do with telecoms to communicate with family, friends, suppliers and customers. And aviation grows ever more popular for longer journeys. With no change in the economic environment, aviation's use of oil will increase more rapidly than any other use.

Biofuels, coal[39] and natural gas can fuel our flying habit, but we don't have enough to keep all our planes in the air. Aircraft manufacturers reduce fuel consumption with each new generation of aircraft. They replace metals with lighter, stronger composite materials, improve aerodynamics, and increase the fuel efficiency of the engines. Airlines are reducing weight, making routes more efficient and employing new landing procedures to reduce fuel costs. Still, we could use some alternatives to traditional flight.

Unlimited solar radiation strikes planes flying above the clouds, during the day. Bertrand Piccard and André Borschberg built Solar Impulse to show we can produce and store enough electricity during daytime flights to fly through the night. In July 2010 they did it. Two years later they achieved intercontinental flight and, on every night of an eight-day return journey from Switzerland to Morocco, Solar Impulse landed with full batteries. During 2015 they went one big step further and set off around the world. The trip is scheduled to last two years, taking 14 legs and flying for 500 hours. This audacious aircraft might be as crucial as the Wright Brothers' Flyer in proving (solar-) powered flight is possible.

Solar Impulse uses batteries to store electricity, and Airbus Group believes advances in batteries and materials could lift a fully electric aircraft into the air by 2017. There is nothing to say that all future aircraft

---

38 Light railways – why let intelligibility get in the way of a good allusion?

39 No, not steam-powered aeroplanes. Do you remember gasification?

will be solar-powered; indeed, powered flight could be achieved through any number of means. Multinational projects such as Hakari and Hexafly International are examining the potential of using technologies such as hydrogen-powered rockets and ramjets to achieve hypersonic flight.[40] As they expand their know-how, they will build aircraft that use much less oil.

Hydrogen doesn't power just rockets. Historically, and recently, aeroplanes from Boeing's Phantom Eye and the B-57 (both demonstrators) to Tupolev's 155 (later converted to LNG) have used hydrogen. These aeroplanes have not made commercial flights; and when we think of hydrogen and flight, we are more likely to imagine the skies over cities filled with airships. Airships do exist outside the ambitions of the Thirties, scenes from science fiction movies and blimps carrying advertising over sporting events. To make the most of modern materials and design techniques, companies are developing airships to realise the potential of lighter-than-air travel. Modern airships use helium to provide lift and the Russians have launched an airship to transport people and equipment to the Arctic (to explore for fossil fuels). Closer to home, the pilot – and lead singer of Iron Maiden – Bruce Dickinson is one of many investors in the world's largest aircraft. The Airlander10 airship hails from Bedfordshire and has the potential to spawn 1,000 even larger commercial vessels.

Boats and ships

Maritime trade uses little oil for each kilo of cargo; however, when we calculate the freight moved, the total is thousands of barrels a day. The oil-based fuel of choice for shipping is marine bunker (also just called bunker). When we ask, later, "What has the oil industry ever done for us?", you will see that even the best refineries end up with thick, dirty hydrocarbons that do not readily convert to lighter, cleaner grades of fuel. Shipping companies buy some of these hydrocarbons, with the chemicals they hold (eg sulphur), to burn at sea. The volume of bunker consumed each year is enormous. The CO_2-equivalent emissions of greenhouse gases from shipping for 2012 equalled 796 million tonnes or

---

40 Hypersonic is even faster than supersonic. Most of the concepts under assessment will fly at (at least) twice the speed of Concorde.

2.2% of total global emissions.[41] By way of comparison, shipping produced about the same amount of carbon dioxide as two whole countries: Germany, the sixth-largest producer of greenhouse gas, and Austria, the 53rd-largest producer.

The relatively low cost of bunker keeps shipping dependent on oil. It costs about $250 per tonne[42] including local taxes and duty. At these prices, alternatives struggle to compete. For the oil companies, selling bunker helps keep down the price of other distillates. Bunker was a common, cheap way of heating buildings until concern about pollution drove out most land-based uses. Top of the offender list is smog, which comes from fuels heavy in nitrogen.

You might think the same concerns would apply at sea, but national interests stop at borders or, in this case, international waters. Pressure on shipping businesses comes from international bodies and customers. The two prominent international organisations are the United Nations International Maritime Organization (IMO) and the European Union. The IMO is searching for technical and market-based ways of reducing shipping pollution. On 1 January 2013 two new rules started the ball rolling. All new ships must have an Energy Efficiency Design Index (EEDI), and all ships must have a Ship Energy Efficiency Management Plan, with a few exclusions. Taking a different tack, the European Union is developing laws demanding any large ship leaving, entering or sailing between European ports report its carbon dioxide emissions. With a fair wind, the proposals will become national legislation and be put in action by 1 January 2018.

Even these new regulations leave us with high marine oil consumption for some time to come. Thankfully, good news sails on the horizon.

> Some shipping companies are moving away from bunker. We can only speculate whether their motivation is reputation, economics or peer pressure, so we won't. In common with other vehicles moving away from oil, natural gas is the first port of call (oops)

---

41 Each litre of bunker has less energy than other fuels, so we cannot say carbon dioxide = oil.

42 The density of marine bunker varies, so it is sold by weight, not volume. In comparison, oil at $50 a barrel would cost about $380 per tonne before local taxes and duty, whereas a tonne of diesel bought on the UK high street at £1.08 a litre would not give you much change out of $2,000.

and **LNG-powered ships** are on sail (sorry). Closer to the drawing board, designers and operators are testing LPG and hydrogen-powered ships.

Another exciting advance is the introduction of **hydrogen fuel cells** to marine craft.

Without changing fuel, Maersk has introduced **emerging technologies** on its new class of vessel, the Triple-E container ship. These measures reduce the oil footprint of each container by 50%.

On a different heading, since December 2009 a traditional-looking clipper called the Tres Hombres has transported goods across the Atlantic in 45 days (weather allowing). From this point, **sail-powered container ships** are but a step away, and Fairtransport, which runs the Tres Hombres, is investing in the Ecoliner, designed by Dutch naval architects Dykstra. The ship is a hybrid that uses sail to maximise fuel efficiency and an electric motor to achieve delivery commitments.

Solar power drives some small boats, and could reduce oil use on larger vessels.

Imagine a kite that looks like a **parascending wing** helping engines power container ships. A company sells such a thing. All large ships, be they cargo, tankers or cruise liners, could reduce fuel consumption in this or similar ways.

When we talk about power for shipping, we must revisit **the nuclear option**. Three classes of marine vessels successfully use nuclear fission. Submarines are the most familiar, and icebreakers (mainly Russian) and aircraft carriers (mostly American) boost the numbers. For commercial craft the story of nuclear is still in its infancy; however, industry specialists Lloyd's Register told Marine Log they expect to *"see nuclear ships on specific trade routes sooner than many people currently anticipate"*, and a review of academic papers found nuclear-powered ships are feasible. But the same study concluded more work on technical and regulatory aspects is required to make them viable. The ships that lend themselves to nuclear power are large bulk carriers with regular routes (eg oil tankers, cruise ships and tugs), and bulk carriers whose cargo makes speed essential (eg those carrying fresh fruit).

Getting it all in place

Where are less-oily investments in transport happening? Governments, local and national, run public transport or delegate responsibility to third parties. Daily commutes, business travel and holidays are the biggest uses of public transport, but there are notable others. In 2012 London hosted the Summer Olympic and Paralympic Games. The events saw more than 10,500 athletes, 21,000 members of the press, 8,000 officials and 200,000 employees descend on London for more than three weeks of sport. During the games, spectators made 20 million journeys. A goal of the hosts of the 2012 Olympics and Paralympics was to regenerate East London. Transport improvements made the events and regeneration of the area possible. Permanent and temporary transport improvements included:

Upgrades to the Docklands Light Railway

Extension of the overground East London Line

Upgrades to the **North London** line

Introduction of the **Olympic Javelin** service

Games lanes on roads

Tube upgrades to the Northern, Central and Jubilee lines

Creation of the **Emirates Air Line** (a cable car across the Thames – and a top pun)

River services

Walking and cycling schemes

The Olympic legacy will expand the Underground, standardise responsibilities for suburban rail services, improve Lea Valley services and put in place new Thames crossings (tunnels, river services and bridges).

Similarly, the Brazilian government had big plans for transport improvements for the 2014 World Cup and the 2016 Olympics. The summer of 2013 saw riots in Brazil sparked by dissatisfaction with transport, and soon afterwards the government launched additional projects to improve the lot of the travelling public. However, the transport projects were dropped as they would not be delivered in time to greet football fans travelling to and within Brazil for a month of competition in 12 host cities. Disappointingly, transport links for the 2016 Olympics seem to be based on existing infrastructure.

But do we need a major sporting event or civil unrest to justify the investment in improved, sustainable transport? Cities around the world have steamed ahead without extreme motivation:

> **Tyne and Wear** – The County Council have a good strategy for transport in the region. Included in their plans are electric cars, with funded schemes to bring greater numbers into the area and introduce charging points across the North East. Newcastle City Council is also a member of a car club and encourages new clubs throughout the city.

> **Bristol** – From 2008-2011 Bristol City Council ran a project to make Bristol into a 'Cycling City'. The project enjoyed much success and its work continues. It attracted international interest and a visit from the Major of Grimstad, Norway, who wants to set up a similar scheme in that equally hilly city. To support people who cannot jump on a bike or have journeys unsuitable for cycling, Bristol is improving public transport. In 2010 the City Council, along with councils from Bath, Somerset and Gloucester, agreed on a 15-year plan to improve public transport in the area. Rapid Bus Transit, metro services and heavy rail are in the plan and have funds, but the most interesting system is the Freight Consolidation Scheme. The scheme serves 50 retailers in a city-centre shopping centre. In other cities each shop, café and restaurant manages its own deliveries. In Bristol, goods arrive at a central depot, where staff group them. Then they are delivered by electric vans, reducing related city-centre pollution to zero. If you live in or visit city centres, you may agree the Bristol scheme is the start of a good thing.

> **Daventry** – Siting the UK's first bio-LNG station outside Daventry is a matter of logistics, rather than concerted action by the town. However, the town is innovative and has explored PRT and GRT and hopes to become a test bed for autonomous cars. Together these advances show we need silver buckshot to build alternative transport.

Bogota – In 2000 Bogota's president launched a project to remove thousands of small private buses from its busy streets and replace them with Bus Rapid Transport (BRT). The improvements have reduced pollution, congestion and traffic deaths. It is the only large transport system approved by the United Nations to sell carbon credits. Bogota has raised more than $200 million by helping others offset their emissions. However, the system was a victim of its own success and overcrowding spurred rioting in 2010. Bogota is now improving service and reducing fares, and other countries are following in its tracks.

Seoul – Another congested city takes a carrot (improved public transport)-and-carrot approach. The second carrot comes in a voluntary system in which drivers keep their car off the road one day a week. Members of the scheme get a host of benefits, from cheaper tolls to free car washes, and local businesses benefit from supplying the scheme. And the city benefits from less pollution and congestion, and reduced costs. New Zealand ran a similar project as part of its 'Think Big' campaign in response to the second oil crisis mentioned in "What's it oil about?

Philadelphia – The South Eastern Pennsylvania Transit Authority (SEPTA) runs public transport in and around Philadelphia. It runs a programme called SEPTAinability. The scheme has three key measures: People (social), Planet (environmental) and Prosperity (economic). Since the start of the programme, use of public transport has increased, and the energy and water used for each passenger mile travelled have decreased. Improved access to work and public facilities improves living standards for poor communities. Commuters in Greater Philadelphia spend 23 million fewer hours on the road – an estimated saving of $473 million of GDP, measured in 2007. The transport company delivered the improvements with running cost increases 5% lower than the industry average.

Emeryville, California – Like other small towns, Emeryville faced the loss of employment to big cities. Since it introduced the Emery Go-Round (you have to admire their style) it has drawn in corporate employers and revitalised employment. The aforementioned transport is a fleet of buses first funded jointly by local government and businesses; local businesses now support the entire scheme. In 18 years, passenger numbers have grown from 300 a day to 30,000. Now, finding a way of keeping up with increasing rider numbers without increasing fares is a challenge the city and residents are facing together. Mountain View, the city at the heart of Silicon Valley, is introducing a similar scheme.

Despite those and other examples, most of us still hop in a car. Our cars are our pride and joy; they are the biggest or second-biggest purchase most of us will make and provide the freedom and flexibility we crave. To reduce the oil cost of transport at a personal level, we need to change to smaller cars, drive more carefully, let the car drive for us, think about the need (and mode) for our journey and change fuels. Perhaps you could put off a trip to the supermarket for a couple of days, jump on a bus, look into LNG, or use a socket in your garage to charge an electric car?

But cars are not the only way we use oil to travel. We need oil to build rail tracks and roads. We need oil to build power stations and factories to produce liquid fuel from waste. We will use oil to build charging points. Oil remains the most popular source of hydrocarbons for the plastics we need to make lightweight, safe vehicles and the kites to pull ships. With so many oil challenges and solutions already in place, no one can accurately predict the future of travel.

One thing is sure, though: we have a long way to go.

Chapter two

Oiling the wheels of industry

ALL THOSE MAGNIFICENT ALTERNATIVES to oil-based travel need oil in another way: the oil we use to make them. In a sidebar, we will look at the oil employed in the production of three materials. Limestone – as it forms the hardcore that virtually every road is built upon and concrete for every other form of transport. Iron and steel – as the basis for most vehicles. And copper – after all, those electric cars are going to need a lot of that conductive material. As a special bonus, there are also a few words about our direct use of oil and chemicals made from oil. These subjects are bookended with a quick look at the history and economic importance of manufacturing, and a section showing how industry is tackling its oil costs.

We human beings are a creative, resourceful bunch. The ancients shaped flint, cured leathers, crafted tools and made decorative items; with time, our ability has grown. Making stuff defines our development so well that we use the names of our materials – stone, bronze and iron – to mark out our prehistoric achievements. Various societies developed similar technologies. They used clays and leather for carrying liquids; stone and metal tools for cutting; cooking to make food safe, edible and more enjoyable; and construction to create shelter.

From the products they left behind, we see that crafted products varied across different societies, but they used the same sources of energy. Human beings worked hard, and animals worked even harder. For millennia we burnt wood, animal fats and fossil fuels. Then the industrial revolution harnessed the power of water and deep mined coal to make high-volume, low-cost production possible. The principles of mass production came before the modern oil industry, and then our sticky friend fuelled manufacturing and economic growth in the 20th century.

Making stuff brings satisfaction and makes the economy go round. If you take flour, eggs, butter and sugar, mix them up and bake the mix, you will have a cake. That is the satisfying bit. In making a cake we add value, meaning people will pay more for it than the raw ingredients (including energy). Adding value is the economic part.

To understand oil in manufacturing, we must look at the entire 'value stream'. This disliked piece of jargon describes the sequence of tasks that add value to materials. From start to finish, all material passes through four main steps. The following uses oil as an example (you will get all the details in "What has the oil industry ever done for us?").

1. Economists call producing raw material, be that fish or stone, *primary production*. Drilling and extraction are the primary production phase for oil-based goods.

2. Because primary production rarely extracts purely the stuff we want, we use *separation* methods, like distillation, to isolate valuable substances.

3. *Processing* converts raw material into something used by manufacturers. For instance, cracking heavy hydrocarbons makes the ingredients for petrol. Some production techniques combine separation and processing. A good example is wheat production, where a single grinding operation removes the wheat kernel from the husk (separation) and reduces it to flour (processing).

4. Now, in *manufacturing*, something useful is formed. This stage uses techniques from moulding and machining to assembly or, for petrol, blending hydrocarbons.

Everything we use passes through several value streams. For example, the primary production of petrol includes mining for zeolite. Some materials, such as plastics and metals, pass through the four stages several times before they are ready for sale. Oil removes human constraints from the value stream, increases the amount produced and reduces costs.

Here is another of those interesting diversions: This 'sidebar' explains the work that goes into producing three key materials and shows how oil-based products are used in industry and at home.

Limestone, also known as calcium carbonate

Throughout the value stream, we use the energy of oil to do more than we can with human or animal power. In the Stone Age, we dug stone by hand, and transport relied on the lifting and carrying strength of people and our various four-legged friends. In the late Stone Age, we invented the wheel. With it, we moved more quarried products farther. The Bronze and Iron Ages brought metal tools. Quarrying served local markets and local needs, with one or two remarkable exceptions like the stones of Stonehenge. Transport experienced a radical change in the late 18th century and canals and railways allowed quarries to sell their product in far-off places. But even with the use of metal tools and dynamite, quarries had to employ people to collect the rock – that is, until we could power machinery.

Powered machines let us extract industrial volumes of rocks from the Earth. In 2013 the UK quarried over 60 million tonnes of limestone, dolomite and chalk. To help you understand the scale of quarrying, here are some comparisons. You would need 332,000 blue whales to balance the scales, which is 33 times as many as are alive. You could build, fit out and load 1,263 Titanics. Six teaspoons of neutron star would fight in the same class. In 2012 all humanity weighed 287 million tonnes, only five times as much as UK limestone, dolomite and chalk. We quarry a few million tonnes of other stones too, from sandstone to shale. Don't forget, that's just in the UK. In 2014 the United States produced 1.26 billion tons of crushed stone, chiefly limestone and dolomite (and several tablespoons of neutron star).

The commonly quarried stones formed in the distant past from the skeletons and shells of marine creatures. Chalk and limestone consist of calcium carbonate, in different forms. Dolomite contains magnesium and its scientific name is calcium magnesium carbonate. We use dolomite as an ingredient in concrete, as a source of magnesium and to make magnesium oxide (used in cement). Magnesium is the third most common construction metal (following iron and aluminium), but calcium carbonate has more uses than dolomite.

Of the quarried stones, we use limestone most widely. It packs out toothpaste and the foundations of roads and is the main ingredient in cement. In 2014 the global cement industry churned out 4.2 million tonnes of the stuff. Cement is the dry powder used to make mortar and concrete. We make cement from limestone, other ingredients and heaps of energy. Indeed, it is one of the most energy-intensive materials to produce. Limestone and clays travel through a furnace to remove water and force changes in their chemistry. They form small hard pellets called clinker. Grinding clinker produces cement. The makers of concrete add other ingredients, including fly ash; aggregates like sand; and ground, granulated blast furnace slag, to change its qualities. For example, gypsum alters the time concrete takes to harden.

People who only need a little concrete for DIY buy the powder, mix it with water and pour it into place. For larger jobs, the concrete is precast or mixed before travelling to the building site. Concrete alone does not make an ideal building material because it breaks under load. Should you add iron or steel, you have a match made in architectural heaven. In the UK we build steel frames and add preformed concrete afterwards. Other countries prefer reinforced concrete, made offsite or poured in place. Moulds hold the metal and give the concrete its form.

While we have used concrete for millennia, it is oil that makes high-volume production possible. In turn, all that concrete shapes our world.

Iron (and steel)

The top ten elements in the universe include iron. It holds the No. 4 spot on Earth and sits at No. 1 (80%) in the core of our planet. Iron combines readily with other elements and makes distinctively coloured ores. These include magnetite or lodestone (black with a grey/ brown tinge), haematite (metallic grey to earthy red), goethite (yellow to reddish dark brown), limonite (yellow/ brown), and siderite (most colours except blue).

People used iron from meteorites before the Iron Age and, as it was rare, they valued it highly. In Mesopotamia, amutum (meteoric iron) was worth 60 times its weight in silver. Most regions came into their Iron Age when the bloomery method of smelting iron ore arrived. China skipped directly to the blast furnace. Bloomeries produce a bloom, or spongy mass of pure iron, holding impurities. The iron maker beats it to remove the impurities in a method called wringing; it results in *wrought iron. Iron products that survived the Iron Age have more carbon than wrought iron, which is almost pure iron. More carbon results in mild steel. (We could call the Iron Age the mild steel age, but it doesn't have quite the same ring.) Skilled blacksmiths could put a fine, sharp edge on steel. With quenching and slow cooling, they made the great swords of history and legend: The Japanese Katana and European Longsword – and Excalibur.*

With more carbon we get cast iron (sometimes called pig iron), which Europeans started to make in the 12th century with the blast

Iron and steel primer

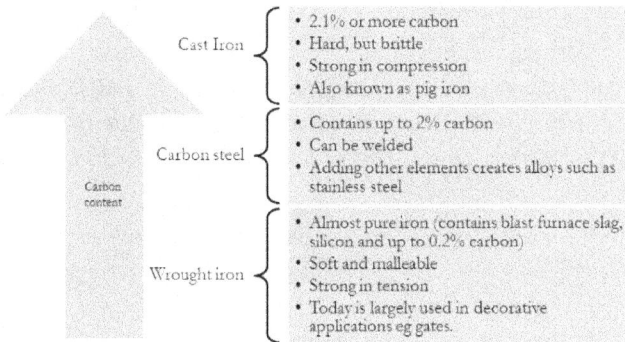

Figure 2: Iron and steel primer

furnace (over a thousand years after the Chinese). The blast furnace does what it says on the tin: Air blasts through heated layers of iron ore, charcoal, more recently coke, and flux. Carbon in the charcoal and coke reduces the melting temperature of the iron and the flux removes impurities.

Wrought iron is soft and cast iron is brittle, so neither is suitable for many of the ways we use metal. Motivated to find a better material, many entrepreneurial engineers and scientists sought a way of making high volumes of steel. The carbon content of steel lies between wrought and cast iron, this gives us three basic methods for making steel:

➢ *Add carbon to wrought iron.*

➢ *Take carbon away from cast iron.*

➢ *Blend cast and wrought iron.*

59

The history of steel has some interesting stops on the long and complicated journey to modern methods. For example, puddling (a great name for an industrial process), which makes wrought iron out of cast iron. The equally strangely-named cementation process then turns the malleable iron into steel. These early techniques were time-consuming and costly.

Then, in the mid-19th century, scientists in the UK and United States discovered that pushing air through cast iron forced the carbon to react with oxygen. The British scientist, Bessemer, won the name, but the American proved he was first and got the money. After that, German scientists developed a more controllable method called the Open Hearth Process.

Now steel was available at low prices. Steel rails replaced less-durable tracks made from wrought iron; steel frames made possible the first skyscrapers; and ship makers showed that, with the right design, metal can float on water. Andrew Carnegie made his money by making cheaper, better steel.

Today, the biggest difference between the methods of the 19th century and modern steel manufacturing is the use of oxygen instead of air to convert cast iron. However, the electric arc furnace has now made it practical to recycle steel. Recycling massively reduces the energy used to make steel – recycling uses 6 to 15 megajoules (MJ) per kilo while making it from iron uses 20 to 50MJ per kilo. And the latter number does not include the additional energy needed to get from ore to iron (20 to 25MJ per kilo).[43]

Stepping back, you cannot throw iron ore into a furnace and hope for the best. Iron mines (typically open-cast) produce various rocks containing around 60% iron ore and other minerals like sand and phosphorous compounds. Ancient Arabic engineers used wind and water to power gristmills to crush iron-bearing rocks. Grinding increased the surface area and made their bloomeries more effective. Now oil has increased the volume of rock we pulverise. Iron producers want the iron ore, the whole iron ore and nothing but the iron ore. In

---

43 Say you recycle and save 35MJ of energy per kilo of steel. What could you do with it? Well, the average UK home uses 16.8MJ a day to heat water. Assuming all the steel made in the world in 2014 was made from scratch, and assuming we replaced all of it with recycled old steel, then we could heat 3.5 billion UK homes. And that, as they say, is enough.

common with oil, better sources of iron cannot keep up with our needs, so mining companies have to remove increasing amounts of less-useful minerals from the ore.

We can give no points for working out that giant magnets make separating the magnetic ores easy. Non-magnetic ores need more cunning. Iron ores have a higher density than the rocks extracted with them – a difference which makes separation possible when magnetism will not work. Blending the ores with liquid and changing the density of the mix allows the iron compounds to fall while other materials float.

Now we have iron ore, we face many processing choices. The one we use depends on the customer, but every steel producer insists on low levels of phosphorus.

When we had enough low-phosphorus iron ores, miners mixed high- and low-phosphorus ores to control levels of contamination. Now they roast the ore to remove phosphorus, but that leaves other unwanted elements, as well as increasing the oil cost of making steel.

We have to get rid of those other impurities, and fluxes do the trick during blasting. (Time-travel fans will note they have no relationship with the flux capacitor.) The most common flux is calcium carbonate, which combines with silica. The resulting chemical, calcium silicate, is blast furnace slag. Chalk and limestone also remove phosphorus, and a manganese compound called spiegeleisen removes oxygen.

All that effort still only makes carbon steel or mild steel, which is not good enough for many jobs. To improve its properties, including resistance to rust, other chemicals are added to make alloy steels, which include stainless steel.

Steel production is one of the largest and most oil-hungry industries. But without it, the global economy would not run. Here is a visual metaphor: imagine a giant doughnut. Bigger. This doughnut can wrap the Earth at the Equator, and would tower over most of us as it stands 2.6 metres tall. We could make such a doughnut with the steel made in 2014. Instead, 40% of the world's steel goes into construction and the rest goes to industrial equipment, infrastructure (eg railway tracks), and metal goods including packaging.

To make useful products out of iron and steel we form them with:

➢ *Forging – bashing steel or wrought iron, hot or cold, into shape. Cold forging results in a stronger metal.*

➢ Casting – *pouring molten metal into a mould.*

➢ Rolling – *passing cast ingots of steel through rollers to produce sheets of steel.*

➢ Extrusion – *forcing metal through a mould to produce long items like pipe, bar and wire.*

➢ Machining – *the two most common machining techniques are turning and milling. Turning introduces a stationary tool to a spinning part. Think about making an ornate chair leg, or the spindle on a balustrade. During milling, the part stays still and the tool spins. If you have seen a circular pattern on a metal item, you have probably seen the evidence of milling.*

Copper

Copper was one of the first metals we used. In the same way that they used iron that fell from the sky, our ancestors found and used meteoritic copper. While they used it (and bronze[44]) to make tools, weapons and pretty things, we tend to use it in more everyday applications. It conducts electricity well, which accounts for its primary role in electricity transport, and other electrical and electronic jobs. It is a common plumbing material and makes fastenings, eg nuts, bolts, screws and washers. Today, copper producers mine and smelt the ore – copper sulphide. More processes remove impurities and make copper alloys.

So far, it sounds like steel production, but we do not have to mine copper. When you read about fracking in "What has the oil industry ever done for us?", you may notice some similarities with in situ leaching. In the latter, operators drill a borehole and crack the rocks with explosives or water, then pump in a solvent. The liquid dissolves the copper and leaves through another borehole. The pure copper is 'won' from the liquid with solvent extraction and electrowinning.

A small proportion of copper goes to alloys (brass and bronze), and more than 90% remains pure to make electrical and plumbing parts. The key manufacturing processes for copper are casting, rolling and drawing. Drawing pulls the copper into wires.

Copper is quite common (No. 29 in the Earth's core), but its ores exist in low concentrations (about 0.6%), so it is expensive to produce.

---

44 Bronze is an alloy (mix) of copper and tin. Brass is an alloy of copper and zinc. I remember which is which as there is one 'z' in both sentences.

High production costs make copper the third most recycled metal (after aluminium and iron) and lead to theft from roofs, electricity installations and railway signals.

Other uses of oil in industry

About 10% of the oil we use each year is not used for energy. Some uses are obvious – we are going to meet plastics in a second, and you travel on oil-based roads and pavements – however, a tiny proportion is used as lubricants. Maybe you have used grease to reduce friction or to stop items sticking together. Well, lubricants have other uses. They:

➢ *Form seals to prevent the escape of gases (grease prevents the smell of the sewers escaping manhole covers).*

➢ *Carry away dirt and debris (cleaning is one job of engine oil, which picks up dirt and bits of metal – so do your car a favour and make sure you change your engine oil regularly (and do the environment a favour and dispose of the oil responsibly)).*

➢ *Protect against corrosion (hence smearing up your bicycle chain).*

➢ *Reduce wear (the primary role of engine oil).*

➢ *Transfer heat (lubricants are used in heat recovery systems and as a coolant).*

➢ *Transfer power in hydraulic systems (brake fluid is a hydrocarbon hydraulic fluid.)*

➢ *Remove waxes (in car engines and fossil fuel power stations, waxes build up on the machinery; lubricating solvents remove the wax).*

We have alternatives to petrochemical lubricants (oils made from plants and animal materials) which can perform the jobs listed above. While they have yet to establish a pedigree, and the earlier forms put pressure on food chains like the first-generation biofuels we met in "Here, there and everywhere", these bio-lubricants will oil the wheels of industry.

Making do with less oil

Climate change, the high price of energy, and the cost and scarcity of raw materials make manufacturing research a profitable venture for industry and academics. You might wonder why anyone would focus on industry when considering how to mitigate climate change. This chart quickly shows that industry is a major contributor of greenhouse gases.

Greenhouse gas emissions by sector

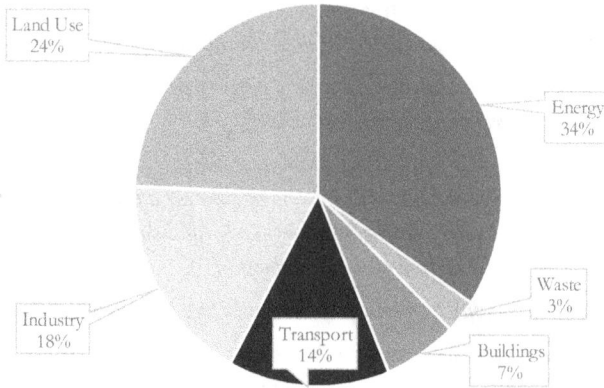

Figure 3: Greenhouse gas emissions by sector

Do you remember that assumption, which we're making, that carbon dioxide = oil? In this chapter, the biggest exception is that the chemistry that turns concrete from sludge into solid blocks emits carbon dioxide. So the assumption does not hold for all manufacturing. Nonetheless, production uses a lot of oil – probably more than transport.

The following gives four areas where industry can reduce its oil costs, followed by examples of how it is doing so. Engineers have developed many, many more.

1. Buildings, pipes and car frames use more steel than necessary.
2. Manufacturing wastes material. Turning and milling from solid block produces thin ribbons or chips of metal called swarf. Stamping sheet metal to create shapes, for example, car doors, leaves an outline of the part in waste. Metal cans are drawn out of flat, circular blanks and drawing damages the edge of the metal, which has to be cut off. Although most manufacturers recycle this kind of waste, the cycle of discarding and recycling increases the oil cost of materials.

3. Manufacturing uses high volumes of water. In the UK, the electrical and other industries directly abstract around 68% of the water we use, which means they take it from rivers, streams, lakes and reservoirs. The water industry takes another 27% and supplies smaller manufacturing businesses.[45] Water use in industry gives a reliable sign of industrial development.[46] Richer countries use more water in industry and poorer countries use more in agriculture. Industry uses hydrocarbon materials and energy to clean water, before and after use. See "Homeward bound" for more information.

4. Historically, other factors drove the cost of production and manufacturers used (cheap) energy to reduce costs.

Matching solution to problem, here are some ways in which manufacturers are addressing these issues:

1. In case studies across five industries, a team of researchers from Cambridge University showed a possible 30% reduction in material through improved design practices; for instance, developing new shapes for architectural beams.

2. Many manufacturers buy blanks, be they sheet metal, bar, forgings or castings, which they machine. At one time it was considered cheaper to buy standard blanks and recycle (or even throw away) the waste. Now many manufacturers favour starting with blanks close in shape to the finished item, calling this Near Net Forming. Where waste is inevitable, it is, at least, recycled. But; if manufacturers use the material as blanks for further manufacturing, then more oil is saved. A company in Stevenage uses sheet scrap to make parts for other firms.

3. Industry can use low-water systems and fix leaks to reduce water costs. Companies save oil and money by saving water.

4. Metal processing companies can update practices to improve energy efficiency. With American government funding, the Ford Motor Company is developing a method to produce pressed

---

45 What happens to the other 5%? Irrigation and other agricultural uses (eg washing cows before they go to the abattoir), fish farming, 'private water supply', and others.

46 Choosing how to describe groups of countries based on their wealth is hard. The terms *developed* and *developing* can be read to mean more than economic development, so I have specified industrial development.

parts with 'features' on both sides. The method removes process steps, reduces energy use, and can help other industries reduce costs. Another example of reducing manufacturing stages may have already occurred to you: Making steel directly from iron ore, not from processed iron, could save a considerable amount of energy. Dutch engineers in the Sixties knew this and took on the challenge, but the external incentives for investment (rising energy prices, climate change) were not recognised until 2004. Now, a project funded by a European Union consortium along with Tata and Rio Tinto has proved a new way of making steel from iron ore and will take the method into high-volume production.

More surprising is the march from steel to lightweight metals such as titanium and magnesium. The US Departments of both Energy and Defense fund projects researching methods of reducing the (energy) cost of extracting and recycling these metals. Their vision is a world built on light, corrosion-resistant metals. In such a world we would need much less energy, would produce less waste and would be able to reuse and recycle almost endlessly.

3D printing builds up layers of material to form a solid object. Spraying material from inkjet heads and laying down plastic fibres are clearly related to printing (and making coil pots). However, there are lots of techniques for building up layers of material that do not have obvious analogies. Pulsed laser deposition uses lasers to evaporate material, which condenses on to a pattern and stereolithography (easy for *you* to say) uses a laser to solidify layers of liquid plastic. The technology seems to know no bounds and every day exciting new 3D printing methods are announced, including cheaper ways of printing metals, printing food and, yes, body cells. 3D printing is set to reduce the oil footprint of forming. These techniques are particularly exciting, as they change the relationship between manufacturer and consumer. For example, public libraries in the US now have 3D printers, and an online retailer is trialling 3D print trucks.

We should also note the increasing use of nanotechnology and automation to reduce energy costs.

You might ask who is paying for all this research and development. The UK government and the EU give financial incentives to the companies developing advanced technologies, and help make technologies developed in universities commercially viable. They intervene because introducing new technologies is an expensive, slow business.

Let's go back to steel. If we move to Near Net Forming or bespoke structural beams, then manufacturers will have to change production more often. These businesses have spent decades standardising production to reduce the costs associated with having a broad and varied product line. To become good at variable manufacture, industry needs money for research. It has to train its people and provide proof of quality to customers and legislators. New machines, dies, moulds and tooling all carry tangible costs. Building and refurbishment costs soar. When accountants calculate the cost of products, they have fewer items over which to spread fixed costs. Finally, unless they are sitting on bags of cash, companies have to raise finance to pay for changes. If we want steel – or any other material – with less embedded energy, then we have to pay for the change. We either pay for more expensive products, pay taxes that are used to support innovation, or pay for imports from countries whose governments support projects like those described a couple of paragraphs ago.

When we look at energy in a couple of chapters, we must remember the needs of industry. Electric arc furnaces do not plug into a 230 V mains socket; they need thousands of volts, and that electricity supply needs its own infrastructure to remain safe. Businesses need to focus on what they do and managing electricity lies outside most skillsets. So, industrial park owners often manage the electricity supply for their tenants.

To find an outstanding example we will travel to Güssing, Austria. Over 20 years the community turned around its economy by reducing its dependence on imported fossil fuels. As a first step, they reduced the town's energy use by 50%. They achieved that goal through a programme of construction and refurbishment. The town then built 27 power plants, including two gasification plants (converting biomass into methane). The larger plant, in Güssing, takes wood from a local forest to supply two megawatts (MW) of electricity and 4.5MW heat. Nearby, in Strem, a biogas generator contributes 0.5MW electricity and 0.5MW heat, using waste from food crops. The investment in low-cost, reliable energy has drawn 50 enterprises to the region. They are largely the high-energy industries we have been discussing. Now, 21,000 people visit Güssing every year to see how the town achieved its transformation.[47]

You might welcome mass production as the means of your consumer lifestyle, or, alternatively, rue the day the first manufactories exploited the power of water. Whatever you feel, we must accept it as the reason we can use oil. Artisans could not refine oil in big enough volumes to keep people on the road, and making plastics without oil would be akin to alchemy. Conversely, if we can't depend on oil, we will have less manufacturing and the basis for a significant chunk of the UK economy will disappear. If we are to continue to enjoy the benefits of manufacturing, then we need to adopt alternative ways to oil the wheels of industry.

---

47 Güssing is far from being the only industrial body moving from fossil fuels. In the Sources you will find a link to a short video about a US metal works that is entirely off-grid.

Chapter three

Plastic dolls

PLASTICS MAKE OUR EASY, FAST-PACED lifestyles possible. From nylon clothing to the durable plastics used in place of metals, they are a cheap, versatile alternative to traditional materials. From the shiny stuff housing electronics to less obvious stuff improving the properties of concrete, plastics help innovation. They give us individuality while following fashion, and they give us belongings that are valuable one day and thrown away the next. The thing is, the raw material for most plastics is oil. In this chapter, we will see how plastics are made and formed. We will also meet the major plastic families. Before that, we shall have a little history and science, and after, we will review five Rs: replace, reduce, reuse, repair and recycle.

The history of plastics

Oil-based plastics are, of course, modern, but plants and animals produce hydrocarbons that we have used for thousands of years. The Mayans used latex (natural rubber) in 1600 BC. We recognised and valued the plastic properties of animal horn, long before we invented petrochemical plastics. We used it to make:

> "…beakers, buckles, combs and buttons. It can also be pressed into thin translucent sheets that can be used for windows and lanterns."

> *—The Worshipful Company of Horners (earliest*
> *extant written reference, 1284)*

Shellac and gutta-percha have been used for items like varnishes and telegraph cables since the 19th century. Waxes sealed documents, made candles, created cosmetics, lubricated machinery, and polished woods and metals for as long. Synthetic plastics have all but replaced these natural materials.

The story of synthetic materials started with the wish to improve natural materials. Around 1820, Thomas Hancock and Charles Macintosh worked out (separately, then jointly) how to produce a thin layer of rubber and glue it to cloth. They created the Mackintosh.[48] Mackintoshes were popular, despite being sticky when warm and brittle in the cold. Perhaps these disadvantages were the reason Mackintosh coats remained the only large-scale use of rubber for some time. In 1838, Nathaniel Hayward patented a way of improving the finish of rubber, using sulphur. He gave the patent to Charles Goodyear, who went on to heat rubber with sulphur and white lead. Goodyear produced a solid which kept the waterproof and elastic properties of rubber and was not sticky. Sadly, Goodyear died in poverty, though the founders of the Goodyear Company recognised and honoured his discovery by naming their company after him. While Goodyear is often credited with the discovery of vulcanisation, Thomas Hancock (the Mac man) took out a British patent one month earlier.

While the people we will meet in "Where did all the oil come from?" were launching the modern oil industry, Alexander Parkes discovered a different kind of plastic. He showed off his invention in 1860 and, modestly, called it Parkesine. It was an early form of celluloid. Alas, Parkes could not turn his scientific find into a prosperous business and instead, John Hyatt built upon his work. Hyatt worked out a robust and profitable method to make celluloid and set up the successful Albany Dental Plate Company in 1870. You may have worked out that he made dental plate blanks. Hyatt anticipated the hot prospects of the new material and set up the Celluloid Manufacturing Company in 1871. Another British inventor, Daniel Spill, had patented the same material and challenged Hyatt's right to make and sell the wonderful new stuff. The New York

---

48 Did you notice the extra 'k'? It seems no one really knows where it came from.

courts declared Parkes the inventor of celluloid but gave the rights to Hyatt.

Vulcanised rubber and celluloid took the world by storm. The uses of vulcanised rubber vary from the familiar waterproof coats and footwear, to tyres, ice hockey pucks, elastic bands, drive belts and seals. In 1855 Goodyear made the first synthetic (proper, round) football. Soon, celluloid plastics copied the items that were previously only available to the wealthy: tortoiseshell combs, ivory billiard balls and porcelain dolls. Equally quickly, inquisitive, innovative minds found new uses. For example, celluloid replaced collodion in photography. The resulting flexible film made moving pictures possible.

The search for synthetic plastics continued. In 1879 Gustave Bouchardat made synthetic rubber, using ingredients taken from rubber. The early 20th century saw rubber made from petrochemicals and now about two-thirds of all rubber production is synthetic[49]. In 1897 Wilhelm Krische combined protein from milk casein with formaldehyde, to create a form of hard casein to replace slate in blackboards. A French company patented this kind of plastic and called it galagith, or milk stone, in 1906. A year later, Leo Bakeland made the first fully synthetic plastic, 'Bakelite'. Bakelite is a phenolic plastic; it is durable, with desirable (technical) properties. Today, we do not use phenolic plastics to make many consumer products. Day-to-day, you are most likely to come across them in your car tyres, or in epoxy resins. However, they are common in electronics, power generation and aerospace.

The modern development history of plastics fills many volumes and, although interesting, does not contribute to our story, so we will stop here.

(some of) The science of plastics

Understanding the science of making plastics is fundamental to understanding how we use oil in a plastic world. The following is a very simple overview.

---

49 In case you are wondering why any rubber is still made from trees, the answer lies in quality. There are some applications for which synthetic rubber is not robust enough, they include surgical gloves, condoms and tyres.

When we distil[50] oil, we get light hydrocarbons, including ethene and propene. These simple substances, which contain two or three carbon atoms, are called monomers. The dictionary definition of monomer is not helpful: 'A molecule that can be bonded to other identical molecules to form a polymer'. Guess how the same book defines polymer: 'A large molecule made up of many identical, smaller, simpler molecules'.

We must not let circular definitions distract us. Let's think of monomers as the stuff we make from oil and polymers as plastics. We have to work hard to make polymers, and the resulting large molecules have names we recognise. They start with 'poly' and may have a trade name, such as Lycra, which is a form of polyurethane-polyurea.[51]

The way monomers connect makes a big difference in the properties of polymers. The most siginificant difference is between plastics that melt (thermoplastics) and plastics that do not (thermosets). In thermosets, the polymer chains link (cross-link).[52]

To make Bakelite (a thermoset) you mix phenol and formaldehyde. You pour the mixture into a mould, in which the ingredients make a new compound. Do not try this at home. The technique is complicated, dangerous and smelly. The starting chemicals are simple, but when they join the result is intricate and has a higher melting temperature, so the mixture solidifies. The new substance degrades at a lower temperature than its melting point, so it does not melt (hence the name thermoset).

Most common plastics are thermoplastics. The makers of thermoplastics slowly heat the monomers to coax them to join. The chemistry is difficult.

To understand the two most common methods of making thermoplastics (condensation and addition), let's imagine a room of people. In the condensation method, everyone holds two buckets of water. To make a chain, they drop one bucket and share the other.

---

50 Distilling oil is one of the separating processes we mentioned in "Oiling the wheels of industry". We will find out much more about this in "What has the oil industry ever done for us?".

51 Its scientific name would wreck the acronymic epithet for 'Middle-Aged Men in Lycra'... so you see why we tend to use trade names when talking about plastics.

52 And one of the most exciting plastic discoveries in recent times is of a thermoset plastic that is easy to make, can be recycled, and, in gel form, can mend itself.

They release water (condensation). If you are ready, we can explore some subtleties. Rather than buckets, the people hold two balloons. One is oxygen, the other hydrogen. The oxygen balloons are larger. A third of people release their big balloon. Two-thirds of our minions let go of their small one. The oxygen and hydrogen combine and make water (condensation). To get really clever, you could give the people other compounds, like alcohol. (Bet you never thought that could be a serious sentence.)

The second common method is addition. Each person clasps their hands together. To form a chain, they let go and grab the hand of two nearby people. Some people hold hands with more than two others and make addition complicated.

Now we have plastics; we add other stuff to make them more valuable:

Colour *is a common additive. Most uncoloured plastics are cloudy white. Pigments, often from oil, lend the bright hues we like in our plastic goods.*

Plasticisers *are much criticised, yet they are an important additive. Where brittleness is a problem, plasticisers are the cure. Below a certain temperature, plastics easily crack. Plasticisers lower this temperature, so they can, for example, facilitate assembly in factories. That 'new-car smell' is plasticisers evaporating from parts that, once fitted, no longer need to be flexible. Large quantities of plasticisers are harmful to health and European law prohibits them over specific levels. The EU has also banned some plasticisers that are still in use in other countries (notably the US). PVC uses a lot of plasticisers; so we do not use it to store food (any more). About 1% of plasticisers are used in soft plastics.*

*Polycarbonate (eg plastic drinks glasses) and epoxy resins contain **BPA** (full name bisphenol A). As with plasticisers, there is potential for BPA to cause harm. Scientific studies show it poses no danger to adults and children below regulated levels, but more research is required to understand its impact on babies.*

*To reduce costs, cheaper **fillers**, like our old friend calcium carbonate, bulk out plastic; indeed, most fillers are minerals. However, wood flour holds the title of the first filler used in*

73

> plastic. *As you might expect, it is ground wood and was employed in Bakelite to improve processing. Other than reducing costs, fillers make plastics lighter, stronger and fire-resistant. Industries from papermaking to concrete production also use fillers, but the plastics industry consumes the lion's share. The filler industry was worth $29.3 billion in 2013 and is expected to keep growing.*

Forming plastics

Making polymers, adding colour and blending them with fillers has been hard work. Yet all we have produced so far are plastic sheets or pellets. Further processing is needed to convert them into items used in industry or sold to consumers. The British Plastics Federation lists 21 forming methods; here are a few of the more common ones:

> **Thermoforming** heats plastic sheets and forces them over a mould. Because it is simple, it is very productive and can churn out thousands of items in an hour. It makes almost every plastic item that has a consistent thickness, including disposable cups, baths, car door panels and packaging.

> **Extrusion** makes tubes. A screw forces melted plastic through a die – just like icing a cake though the die is fancier than a piping nozzle. Extrusion can create the hollow form of a drinking straw or sewage pipe; you won't see that on your favourite baking show.

> **Blow moulding** is similar to glass blowing. Huge machines push air (or other gases) into a slug of plastic held in a mould. Since Nathaniel Wyeth patented the first PET bottle in 1973, the number of bottles made with his method has ballooned to tens of billions.

> If we need a complicated shape, we could adapt the methods above or use simple forming techniques like bending or welding. However, only **injection moulding** can produce complicated, high-quality items cheaply and in high volumes. Hyatt, the guy who made celluloid, designed and built the first injection moulding machine. Injection moulds have two parts: male and female. When connected they form a cavity, into which a large screw forces melted plastic. When the plastic solidifies, the mould comes apart and spits out the moulding.

Tell-tale marks on all but the highest-quality items show the method used to make them. A smooth finish with a dimple on the bottom suggests blow moulding. A join line with an injection point (another dimple) indicates injection moulding. Thermoforming and injection moulding may leave ejection points – small, flat, often circular marks.[53]

After forming, most plastic goods pass through a finishing stage. Blades and razors come together. Fizzy pop and tops complete the bottles. Printer components slot into housings. Ink cartridges are fitted with electronics, filled, and protected with even more plastic parts. We have two main methods for joining plastics. Plastics are difficult to glue, so factories use moulded-in clips. However, because clips harden and become brittle they can break in use, so some plastics are joined by metal screws. When safety is an issue, screws with special heads replace the more familiar cross-headed variety.

The principal families of plastics

Plastic types number in the hundreds of thousands. You may have heard of these well-known families.

PET is a polyester commonly used to make plastic bottles.[54] Polyester was a dead end on the route to replacing silk. If you turn up your nose at the thought of polyester clothing, you should check your labels. Aware of the disadvantages of pure polyester, scientists blended it with cotton to give wrinkle-free material with a natural feel. It also makes toasty clothing like fleeces.

Polyamides resist heat so you will find a version of nylon in many domestic appliances. DuPont, who funded a science department in 1927 to prove or discover new scientific facts, created nylon (launched in 1938). Next, DuPont conducted a campaign to replace women's silk stockings with 'nylons' (repeated after World War II). This concerted drive was a first and led the way for modern industry, filled with strategic marketing and R&D.

---

53 Manufacturers also mark plastic items to record the time and date of production, and to identify the mould tool and plastic used.

54 Strangely, there is no -obvious reference to esters in its chemical name: polyethylene terephthalate.

DuPont did not protect the word 'nylon' (it is not a trademark), nor did they name it after **New** York and **LON**don. The official version of events gives the name as a modification of the phrase 'no run' and says a committee devised it.[55]

Polypropylene has excellent heat properties and includes varieties that are safe for use with food. Therefore, it makes small domestic appliances, eg kettles and toasters. Before a European retailer or food manufacturer uses a plastic, they have to ask its maker for proof it is food-safe. The plastic producer must get the evidence from relevant national testing bodies before its plastics go on sale. In the UK the Food Standards Agency enforces these laws. The plastics in toys and toiletry containers are subject to similar laws.

PVC without plasticisers is called **PVCu**. You might remember its old handle, uPVC. It is rigid and is, therefore, useful for construction, windows, pipes and cladding. With plasticisers, it is flexible; it insulates wires and acts as a substitute for leather and rubber.

Polyethylene, also known as polythene (poly-ethene), makes cheap, throwaway items. It becomes carrier bags, packaging and takeaway containers. It changed the way we view our belongings.

Phenolic plastics descend from Bakelite and share its characteristics. They hold back fire, bind other materials, and make laminates (most commonly with paper or glass), and we can shape them with machines. The brakes and clutch in your car contain phenolic plastics. Art forger Han van Meegeren mixed his paints with phenolic formaldehyde – ie Bakelite – to make his pictures look centuries old. But that is a somewhat specialist use.

Clearly, plastics are important to us. Indeed, it is hard to remember how new they are. An old family story describes a wedding reception held in a house in a village in the North East. The proud parents of the bride had invited many guests and bought a plastic carpet protector for their hallway. Impressed family and friends carefully inched their way down

---

55 I think you can appreciate why the urban myth is more popular.

the hall, trying only to tread on the carpet, rather than spoil its shiny cover. The wedding took place in the Fifties.

In 2012 we used around 1.25 billion barrels of oil to make plastics (4% of global oil production). For comparison, you could fill 85,000 Olympic swimming pools with that oil. You need the same amount of oil again for the energy used in plastic manufacture. Reducing that number may not be as hard as you think; we have many options, as we are about to see.

Replace – Bioplastics, biofillers, and plastics from air

As the story of plastic started with plants, we might find an ending in the same place. We could replace oil-based plastics with bioplastics, sometimes called renewable plastics.[56] We have two sources of feedstock for bioplastics: crops (food) and biomass (food waste, like orange peel, non-food crops like fungi, wood and straw and even carbon's close cousin sulphur), and the biofuel advances we met in "Here, there and everywhere" promise more. Hard-wearing bioplastics are becoming more common in consumer electronics, cars, domestic appliances, soft furnishings, clothes and buildings. The European Bioplastics association believe that production will grow from a 2013 baseline of 1.6 million tonnes to 7.8 million tonnes in 2019.[57]

We know fillers improve the properties and cost of plastics, and that the first filler was wood flour. Today biofillers are big business, even with conventional plastics. They include bamboo, kenaf, corn stover[58], soybean hulls, chicken feathers, and dried grains from alcohol production.

---

56 We must take care: only the *source* is renewable. These plastics have the same chemical composition as their petrochemical counterparts, and responsible disposal remains important.

57 One of these bioplastics could soon grace your living room floor in the form of Lego. The Danish company have announced an investment of $150 million to find a sustainable, safe, reliable alternative to using virgin, oil-based plastic to make their colourful bricks.

58 What on Earth are kenaf and corn stover, you might ask. Kenaf is a member of the hibiscus family, which quickly grows into tall, unbranched plants. It is a traditional material for rope and cloth making. Paper makers use kenaf because it has advantages over wood pulp. Corn stover is the leftover leaves and stalks of maize (sweetcorn).

Scientists have proved the science behind making plastics from carbon dioxide. The work to develop commercial manufacturing process continues, and you could soon see plastics from power station waste in the shops.

Reduce – Giving up the plastic habit

Perhaps the most frustrating use for plastic is packaging. Businesses from supermarkets to car makers have moved from single-use packaging to multi-use crates and containers. Yet the consumer, who has been asked, cajoled, nagged and legislated into giving up single-use carrier bags, receives disposable plastic packaging with almost every item they buy.

In August 2008, Christine Jeavans (a BBC journalist) recorded every piece of plastic she threw away or recycled during the month. Her list totalled 603 items. Other than 120 disposable nappies, most waste came in the form of food or drinks packaging. During September 2008, she tried to buy nothing made of or wrapped in plastic. She could not.

She found some uses of plastic easy to avoid; for example, plastic bags, which she replaced by shopping in a market that used paper bags. Others were more difficult. Biodegradable, disposable nappies still contained plastic, albeit made of corn starch, and dental care without plastic failed miserably. Her family rejected both wooden toothbrushes and homemade toothpaste.

However, she did reduce the overall plastic count to 116. If we all undertake this challenge (and keep it up), we could massively reduce the amount of oil used for plastics.

Reduce – better design

In "Oiling the wheels of industry" we learnt we can reduce the material and energy used in manufacturing by changing design practices. That's a two-edged sword for plastics. On one hand, packaging makers reduced the weight of bottles and films by 21% and 36% respectively in the UK during the Nineties. On the other hand, manufacturers are turning to plastics to reduce the energy costs of manufacturing, transport and use.

Perhaps the biggest opportunity for design is to make plastic goods more durable, repairable or recyclable.

Reduce – the polystyrene story

Expanded polystyrene (EPS) is one of the few plastics we do not recycle. Why? Because we have to wash food off it first. And because it is mainly air, and takes up a disproportionate amount of space in recycling centres and transport. Putting those two reasons together, you can see that recycling EPS is simply not cost effective. So we have two options: throw it in the bin or stop using it. If you are old enough, you might remember that a global burger chain stopped using EPS to pack its eponymous burger and turned instead to paper in 1990. If you are not that old, you may recall that in 2013 New York City banned the use of EPS in all food outlets and as polystyrene packaging peanuts. In both cases, the answer to concerns about the environmental impact of the material was to reduce its use.

Reuse – refill

Reusing containers by refilling them isn't a new idea. Milkmen still deliver milk bottles and take away the empties for cleaning and refilling. Drinks manufacturers place a deposit on their bottles so that bars and consumers return them for reuse. Household chemicals, like window cleaner and hand wash, come in robust refillable packaging and lighter refill packs. Some foodstuffs, eg coffee, come in refill packets too.

The refilling nirvana would be to take our used, cleaned bottles to a shop and fill them there. Some retailers have made a punt at refilling, but even after successful trials the practice has petered out. A local 'health food shop' has a different approach to refilling. The store takes back plastic bottles, cleans them and refills them itself. So all you do is swap bottles, and at a lower price than buying a brand, spanking new one. You can also 'refill' at home. If you buy large bottles of toiletries, eg shampoo, and decant them into smaller (cleaned) bottles, you will save money and reduce your plastic use.

Reuse – give away

Any action that reuses plastics instead of throwing them away is good. Online auction sites, charity shops and 'swishing' events reduce plastic use.

But why give away to strangers? Might we be about to create heritage plastics? Look around your home. What have you inherited? Perhaps jewellery, paintings and other artwork, books, furniture? Unlike our parents and grandparents, we don't own many practical items passed down through the generations, or that you used once and stored for later reuse (except your bread maker). Keeping and reusing plastics is the last word in reducing demand for virgin plastics and, therefore, on oil. Plastic is likely to be expensive for our children – they will appreciate these heirlooms.

Repair

Wouldn't it be nice to be able to fix a broken pair of glasses or a patio chair? Unfortunately, plastics are difficult to glue and sticky tape doesn't work either. We can screw plastics together, but have to design them to take a screw. Otherwise, vibration works the hard metal against the soft plastic and wears away a larger hole. In short, you can't repair plastic without specialist equipment (anyone for welding?), so we have to think about designing and selecting products for durability instead.

Recycle – who uses recycled plastic?

Like anything new, plastic recycling had some issues when it started in earnest, 20 to 30 years ago. The plastics we took to recycling centres sometimes went to overseas waste sites, as the market for the material was tiny. Low use rates were due, in part, to the plastics industry's reluctance to use recycled material, as they could not be sure of its provenance. Today, plastic recycling can make good-quality raw materials, synthetic oil, and energy.

Let's consider the all-pervasive plastic bottle. You'll remember that these bottles are made from PET, a form of polyester. The distinctive shape of PET bottles makes them easy to sort from other plastics, and recycled PET can replace virgin feedstock. Polyester recycling uses the same method that makes the plastic (transesterification), so the recycled material has the same properties as new. Recycling PET uses less energy than making it from oil, so it is cheaper. Using recycled PET is popular; the US National Association for PET Container Resources (Napcor for short) says even more would be used if supplies were available. They are working to make PET recycling better for everyone.

Recycle – improving the quality

We just learned that one of the biggest concerns for people using recycled plastic is contaminated stock. Contamination spoils the properties of the plastic they are making. Asking us to sort plastics has several flaws, so local government and plastics manufacturers have tried to improve our performance. First came the plastic resin code. If you look closely at most plastic items, you will find a number in a triangle somewhere. The number identifies the six most common plastics, and 'others'. However, you rarely see those numbers on recycling information. Next we had unhelpful descriptions, eg 'Yogurt pots okay'. We didn't know whether that extended to other types of containers. Now the recycling powers-that-be seem to have given up on us. Instead, enormous recycling factories separate the plastics, and all we do is split our plastic waste into two neat categories: containers and films. Alas, the latter still go to landfill.[59]

A technology that helps take the onus off us is Near-Infrared Spectroscopy (NIR). This whizz-bang technique uses the unique vibrations of molecules to fingerprint each material. Because of the vibrations, each type of plastic absorbs different wavelengths of light in the infrared spectrum. Cameras measure the unabsorbed light, and – hey presto – we have identified the plastic. Provided it is not black.[60]

Another technology, 'active disassembly', helps recycle electronics, as they contain a mix of materials, some of which are harmful. Active disassembly is a method of designing and making devices that disassemble automatically given the right conditions. One technique relies on materials changing shape as they absorb heat – for example, in an oven. When the parts alter, they spring apart and break up mobile phones and other gadgets. Beyond providing the heat, the oven, with careful design, separates each part. Such disassembly makes recycling safer and more profitable.

---

59 It is possible to recycle most plastic film (anything made from polythene), but only 20% of local authorities support this form of recycling. The government and WRAP are helping councils standardise recycling; hopefully this will both increase the recycling of films and improve communications.

60 You may wonder, then, why some plastic items tend to come in black, like plant pots. The reason is not a strange homage to Henry Ford. It is because they are made of recycled material, and only dark dyes can mask the colours of mixed plastics.

Reducing oil-based plastic use to the essentials would free oil for other purposes, such as building alternative sources of energy, providing medical care and growing food. We will also see strong environmental arguments for wasting less plastic in "Where has all the oil gone?".

However, plastics are an important part of our economy. In the UK, the plastics industry is bigger than automotive and pharmaceutical put together and, in 2011, accounted for 7% of manufacturing. So, using fewer plastics will affect our finances. Turning to bioplastics is similarly difficult, as they increase pressure on food production and need agricultural practices that depend on oil, or require investment to develop new sources.

One plastic is essential to getting through this problem. Our brains are 'plastic'; that's how we learn. The technical term is neuroplasticity. Let's make the most of it.

Chapter four

Taking oil my energy

IF WE USE THAT BROAD DEFINITION of oil from the introduction (nip back if you can't remember), then you will expect energy to be an enormous and fascinating topic. You will not be disappointed. As well as introducing the main forms of energy, in this section, we take a tour through the history of our interaction with electricity. We need to look at the system that gets electricity from producers to us so that we can understand the exciting advantages of a smart grid. Then we will dive into the world of non-carbon electricity production, meeting scientists, businesses and consumers who have started the journey to a low-carbon world. And, of course, we will contemplate energy efficiency. We close with a look at power cuts so that we understand the consequences of failing to improve our electricity supply.

So, let's think about all the types of energy and how they become available to us. Hundreds of thousands of years after the 'Big Bang', matter separated from background radiation. The matter was mainly hydrogen, which gravity pulled into great balls of fire – the first stars. These stars converted the hydrogen to helium and expelled loads of energy. After full and productive lives, they collapsed and exploded.[61] These explosions created supernovas and heavy chemicals like iron and uranium. In our solar system, the heavier materials took ten million years to spin from the centre of a supernova cloud and form the planets. Hydrogen, helium and a smattering of other substances created the nuclear furnace that is our Sun. These events produced all the energy we experience on Earth.

Since its birth, the Sun has converted hydrogen to helium. As a result, it releases heat and light. The heat creates rich and varied weather patterns, especially winds – winds that have so much kinetic energy they overturn cars and knock down buildings. Winds transfer their energy to water and form waves – the waves that beat our shores and are powerful enough to erode solid rock and toss huge ships around like toys. The Sun's heat also causes the oceans to evaporate and wind pushes rain clouds over land. The rain falls and runs into rivers, which flow downstream. When rivers run off cliffs, they release potential energy in the form of noise and by eroding the rocks below. Photosynthesising plants and bacteria use the Sun's light to function and grow.

The Earth's core has been cooling for billions of years but remains hot. Most scientists think the temperature is about 4,000°C though some think it could match the Sun's surface at around 6,000°C. We experience the Earth's heat through geological events such as volcanic eruptions, earthquakes and naturally hot water.

The elements made in supernovas are a mix of stable and radioactive atoms. Radioactivity measures the energy released by atoms as they change. We call this energy radiation or nuclear energy, and it contributes to the heat at the centre of our planet.

The Sun and our little sister planet, the Moon, attract the surface of the Earth, causing tides in rock and water. Tidal energy moves millions of tonnes of material throughout the day and night.

All life on Earth uses all these forms of energy to stay alive, grow and reproduce (yes, including radiation). When creatures die, others eat them. A tiny few have the energy in their bodies preserved in fossil fuels. When

---

61 If they were big enough. This also happens to subsequent generations of stars. Our sun is not so big and will die much less spectacularly, expanding and cooling until it reaches out and engulfs the Earth (as a red dwarf), before collapsing back into a white dwarf. But we have about five billion years to worry about that eventuality.

we burn these fuels, or watch a piece of sodium spark and fizzle in water, we see the release of chemical energy. When lightning strikes a tree, we catch a glimpse of the awesome power of electricity.

Electricity has advantages compared with other forms of energy. It does not cool, slow or quickly lose its power. We can control its dangers, switch it off, transport it through wires and store it in physical containers called batteries. We cannot create or lose energy, but we can convert most types of energy to electricity. This chapter assumes electricity's benefits will make it our principal energy carrier. Changing the basis of the economy from oil will need changes in the way we move and manage energy. Electricity requires fewer adaptations and, as such, is the best horse in the field – notwithstanding silver buckshot (wait for it).

A little bit of history

We are quick to credit James Watt, Michael Faraday, Alessandro Volta and Benjamin Franklin with our understanding of electricity. These scientists, and many others, made great strides in understanding this intangible-yet-valuable, wild-yet-tameable stuff, but the history of humanity and electricity goes back much further. One of the world's largest engineering companies uses the name Thales to express innovation and excellence. The original Thales was a Greek mathematician, philosopher and experimenter. Around 600BC he moved light objects with a rod of charged amber. To create the charge he rubbed the amber with silk. The modern technical term for generating static electricity this way is triboelectric charging; it happens when a charge moves between materials – from silk to amber in Thales' experiment.[62]

Thales' actions are the first recorded examination of static electricity. He named the charge 'electron', the Ancient Greek word for amber.[63] We now understand the electron to be a subatomic particle. Thales also recognised the similarities of static electricity and magnetism. 300 years later, the Chinese started to navigate with magnets on land and, eventually, at sea. Skipping forward several centuries, the benefits of magnetism in navigation and, therefore, in

---

62 We use the same effect to entertain children by sticking balloons to the ceiling after rubbing them (the balloons) on our heads. In that game, the charge travels from hair to balloon.

63 The sticky resin made by trees, not the light between red and green.

exploration and trade, brought it to the notice of the Elizabethan court. There, William Gilbert (court doctor) experimented with magnetism and identified how it differs from static electricity. He published his work in a book called 'De Magnete' in 1600. Gilbert used the word 'electrics' to mean 'like amber'. A little later, Sir Thomas Browne built on that to develop the term 'electricity'.

In the same century, Otto Von Guericke built a machine to make static electricity.[64] Static electricity is less useful than electricity with a current though we do have critical uses for it. Some science experiments use static electricity and, if you have visited a science museum, you may have seen lightning created in a lab. In day-to-day life, photocopying, printing and defibrillators all rely on separating the positive and negative charges in atoms. However, we mainly use the other thing called electricity.

In the early 19th century, Michael Faraday discovered the law of electromagnetic induction and created the first dynamo. He spun a conductor in a magnetic field. Despite words like conductor and magnetic field, it is (in retrospect) a simple principle. Moving copper (the conductor) inside a magnetic field causes a current[65] to flow in the copper. James Clerk Maxwell named the law of electromagnetic induction after Faraday.

Like Rubik and his cube some 122 years later, Anyos Jedlik built the first electricity generator in Hungary (in 1861). Jedlik did not realise he was first to build a generator, so he did not patent his invention. Siemens of Germany (1866) claimed the patent and took credit as the inventor of the electrical generator. Most strangely, the electric motor was invented 30 years earlier, after a long and complicated birth. Electric motors and generators look similar because they use the same principle. Motors use the law of electromagnetic induction, in reverse. Current flows through copper wire and causes it to turn within a magnetic field.

The 19th century was a time of great scientific exploration and engineering achievement:

> *In 1878 **Augustin Bernard Mouchot** made ice with the first solar engine, at the Universal Exposition in Paris.*

---

64 He also built the first vacuum pump and showed that vacuums do not 'suck'.

65 See Appendix Three to learn a little about electrical terms and units.

Mouchot explored solar power because he believed that the coal fueling the Industrial Revolution would run out and he wanted to find an alternative. In his demonstration, he used the Sun's heat to make steam and powered an ice-making machine. The device worked by evaporating a refrigerant – the way gas-powered fridges work today.

*In 1847 **Lord William Armstrong** founded a company to build and sell hydro-powered cranes and used his expertise to be the first to light a home with hydroelectricity. He used light bulbs created by Joseph Swan. Armstrong was the 'Armstrong' in Vickers-Armstrong. He designed the hydraulics powering Tower Bridge in London and the swing bridge that allowed large ships to access his factories on the Tyne in Newcastle-upon-Tyne.*

*In 1878 **King Ludwig of Bavaria** commissioned dynamos and carbon arc lights from Siemens to light the Venus Grotto at the Linderhof Palace.*

*In 1881 the **world's first power station** opened in Godalming in Surrey. Water turned the conductor.*

*In 1882 **Edison** built the first steam-powered power station (Pearl Street Power Station in New York).*

History tells us these innovators backed the wrong horses. In 1895 Nikola Tesla and George Westinghouse built the first Niagara hydropower station. Unlike Siemens, Godalming and Pearl Street, their power plant produced electricity with an alternating current (AC), and electricity knocked the direct use of energy (eg solar) into second place.

Before Niagara, engineers and scientists preferred electricity with a direct current (DC). Even Edison, the great inventor, failed to recognise the inadequacies of DC. During distribution, low-voltage DC quickly loses its oomph and our forebears generated and transported DC electricity at the same low voltages they used. For example, in Godalming, the Direct Current electricity weakened so much that the lights at the end of the town did not shine brightly.

In contrast, we can easily 'step up', or transform, AC. That means increasing the voltage and decreasing current. Transforming improves the efficiency of electricity transport, as higher voltages travel with fewer losses.[66] AC is also easier and, therefore, cheaper to generate.[67]

Edison was not going to give up on DC without a fight. So he employed Arthur Kennelly and Harold Brown to develop the electric chair, to convince people of the dangers of alternating current. At the same time, the state of New York sought a more humane method of execution than hanging, and Edison found a captive market. Brown and Edison coined the word electrocution – a contraction of 'electrical execution'. They did not use it to mean accidental death by electricity. Despite the horror of the electric chair, AC became the current fashion.

In the late 19th century, big was beautiful. Factories replaced cottage industries. New transport systems let businesses trade in greater geographical areas. France, Germany, Spain and Italy were born out of smaller states. Electricity generation followed the trend and became bigger, and its growth depended on fossil fuels.

Bucking the trend, some individuals, communities and businesses took advantage of the energy around them. Some lived in remote locations. Some followed political beliefs. Others did not want to spend money on commercially-produced electricity when they could use 'free' energy. Although their motives varied, they shared technologies dating from before the mass generation of electricity. With their actions, they vindicated the beliefs of Mouchot, Armstrong and Edison, and laid the foundations for distributed (eg home) electricity production.

We are going to see that we have oodles of generation choices, and like any good oodle, they just keep coming. They support both high-capacity production and local requirements, and they can meet our energy needs. People have employed these technologies for decades, insulated from the vagaries of fossil-fuel pricing. They also enjoy greater energy security than the rest of us. For example, Denmark will save $920 million on energy costs by 2020, by converting to a mix of renewable technologies.

---

66 It is not the high voltage that reduces losses. You may have learnt a strange formula at school: P=IV. It means that the amount of power in the electricity is always equal to its current (I) multiplied by its voltage (V). For a fixed amount of power, if we increase the voltage, the current has to drop. This is good, because low currents experience smaller losses.

67 The 'alternating' bit tells us the current switches direction; in the UK it changes 50 times per second. Direct current flows one way.

The country will save this money as part of a plan to have all electricity and heating coming from non-fossil fuels by 2035 and to take the carbon out of industry and transport by 2050. Similarly, France enjoys some of the lowest electricity prices in Europe, thanks to its investment in nuclear power.

While these excellent examples flourish, we who remain dependent on fossil fuels have a lot of work to do. Renewable sources vary with location so technologies that work in the Canaries may be less useful in the United Kingdom. Historically, renewables have enjoyed far smaller levels of investment than fossil fuels and need more money and time to become our sole energy source.

Why non-fossil fuels are not yet ready to be our only source of electricity

The efficiency and effectiveness of energy sources - EROEI and capacity factors

When we think about efficiency, we compare the input with the output. For example, to measure the efficiency of your car, you balance the litres of petrol you put in against the miles you get out. For energy generation we ask how much energy a technology produces when compared to the energy used to make it. The technical name borrows from the financial term 'Return on Investment'. We add the word 'energy' for clarity, and end up with 'Energy Return on Investment (EROI)' or 'Energy Return on Energy Invested (EROEI)'. (There is an actual efficiency measure, too. It compares the energy absorbed to the electricity produced.)

Effectiveness compares the actual outcome with the desired result. Back to your car: If it can only hold enough petrol to drive 50 miles, but you want to go 100, it is not an effective choice of transport (especially if there is no petrol station within 50 miles). Its near relation in energy is capacity factor, which tells us how much energy we get when compared with the designed and tested maximum.

We can use solar to understand how both measures work. Let's start with the capacity factor. The size of solar systems is described by the amount of electricity they can produce, it is normally in kilowatts, and the figure tells us the maximum power the system can produce. As solar systems only operate during the daytime, their capacity factor cannot be higher than 50%. The capacity factor drops further when you take the angle of the sunlight, clouds and other inefficiencies into account. Indeed, the capacity factor for solar power plants is between 10 and 25%, compared to 90% for nuclear.

A low capacity factor is likely to mean a low EROEI. For photovoltaics, a scientific paper published in April 2013 showed solar without storage to have an EROEI of 3.9. The value falls to 1.6 when lithium batteries are added. For nuclear, the figure is 75 (with or without storage). So solar, backed up with batteries, could only produce 1.6 times as much energy as we use to build it. It's worth remembering that EROEI is a new subject, there is considerable debate over the numbers above and, as we improve technologies, both capacity factors and EROEI improve.

We have a choice to make: do we buy these technologies, use them and help their development; do we fund development through other means; or do we stick to fossil fuels?

The space required by energy sources - energy density

Energy density describes the amount of space needed to generate a given amount of energy. Traditional fossil fuel plants, nuclear power stations, conventional hydro and deep geothermal all have very high energy densities; that is, they need little space in which to produce shedloads of energy. Solar and wind both have relatively low energy densities. If you make certain assumptions – eg that solar power has to be built on a large scale – then its low energy density stops us generating enough energy. If instead we think about solar on roofs and built into windows (or even the glass panels on modern office blocks) then energy density is less of an issue. To put it another way, if you can produce enough energy on your premises, then you don't need any form of power station down the road. Of course, there are lots of people who don't have this space (eg in cities) and there will always be cloudy days with no wind. For both reasons, we will need generation techniques that have high energy density and are not weather-dependent. You will find plenty of those in this book.

We have a choice to make: do we tackle these issues constructively, or prevaricate in a state of "it's too hard" and continue to burn oil?

Economics and politics

Uncertainty stops us turning to low- and no-carbon energy sources, and that prevents investments that would make these technologies more viable. For example, our fears about nuclear power have spawned a hundred books and thousands of articles, opinion pieces and debates about safety. Meanwhile, thousands die from fossil-fuel pollution.

We have a choice to make: do we look at the facts and make rational decisions; or do we dither, waiting for a risk-free solution, and default to sticking to fossil fuels?

In addition to the reasons above, we cannot switch entirely to renewables straight away because our electricity system grew around centralised production, which we vary with demand. To use renewables, we will need a better way of managing electricity.

The smart grid

Large power stations sit some distance away from the users of electricity. To connect them we need transmission equipment, typically called a grid. At its simplest, a grid is a power station, a few cables and the end consumers. But life is never simple, so we also need meters, substations, insulation, miles upon miles of cables, the pylons to carry them[68] and commercial agreements to make it all work. Our grid evolved to serve centralised electricity production, and if we designed an electricity grid today, it would have some fundamental differences. The new grid would be smart. It would:

- Accept energy from distributed and variable sources
- Store energy
- Manage all energy, not just electricity
- Manage demand and supply (known as load shifting)
- Cross borders
- Encourage efficiency
- Charge for managing energy, not producing electricity
- Recognise silver buckshot (don't skip ahead)

---

68 Unless we put the cables underground, or perfect the art of broadcasting electricity through the air.

What's more, every dollar invested in a smart grid in the United States returns between $2.80 and $6. For that reason alone, utility companies around the world are making electricity grids smarter.

Accept electricity from distributed and variable sources

For most of the rest of this chapter, we are going to think through many types of renewable or low-carbon ways of producing electricity. Some of these methods could replace traditional power plants (for example, a coal plant could be swopped out for nuclear); others behave differently and take more work. Smart grids can accept electricity from home generation during the day and pay the owner for their contribution. Smart grids can handle the variability that comes with technologies such as solar and wind. Smart grids operate safely and reliably, regardless of the number and variety of such sources.

Store energy

Historically we have not needed to store electricity; at peak times, we just produce more. So storage seems an enormous challenge. But the number of storage options grow every day. At one extreme, pumped hydro schemes use surplus electricity (for example, electricity produced at night) to pump water up hills. When demand peaks, the water flows through turbines[69] to produce the electricity and meet our needs. Similarly, but on a grander scale, artificial islands could act as giant energy stores. These sea-based islands would contain lagoons, which empty using excess electricity from wind and solar power. During peak times, the lagoons fill through turbines and produce electricity. Heavy users of electricity could locate on these islands, reducing transmission losses. No country has started to build such an island, though; for example, changes in the energy market have postponed a proposed 'energy atoll' off the coast of Zeebrugge, which would use similar principles.

At the other end of the storage scale, capacitors (a temporary energy store), charged with otherwise-wasted energy, give a short-term power

---

69 Turbines look like fans and take the energy out of flowing fluids (eg water, steam or air). Taking advantage of the law of electromagnetic induction, we use them to generate electricity by turning conductors within magnetic fields. If you can't quite remember the basics about electromagnetic induction, flick back a few pages for a simple introduction.

boost. For example, they could replace bulky power packs for domestic electric drills and other cordless tools. Big vehicles can use larger capacitors to supply extra energy for starting, braking and lifting heavy loads. Boosting energy with short-term storage reduces the size of their engines and improves overall fuel consumption. Similarly, KERS is familiar to followers of Formula One racing and is now reasonably common in production cars. You may recognise the alternative name, regenerative braking. In one form, a flywheel stores the energy released as the car slows. And in other types the brakes produce electricity to charge a battery. Richard Feynman[70] had the idea of harnessing the kinetic energy (movement) of cars to generate electricity in the Fifties.

Chemical batteries, including the lithium-ion batteries in your gadgets, are improving. Engineers and scientists are reducing the amount of raw materials they use, improving their power and increasing their life. And they are already in use around the British Isles; for example, lighthouses and solar buoys around Scotland and the Isle of Man use lithium-ion batteries. The batteries in these remote locations are bigger than your phone battery and store enough solar electricity in the summer to keep the lights shining throughout the winter. A wind farm in Hemsby, Norfolk uses lithium-ion batteries to store up to 200kW of electricity, which smooths the supply into the grid.

Europe's largest energy storage and control project is taking place in Leighton Buzzard. The project has several goals: to stabilise the grid more effectively than current technology, so that more renewables can connect; to provide load shifting, delay the extension of the grid and postpone the purchase of new transformers and cables. Most tellingly, the oil companies that are turning to solar energy to reduce costs are also investing in batteries to smooth supply. In March 2015 a French power company won an order to supply 44,000 battery systems to store solar energy in Qatar's Dukhan oil field. The energy stored in these batteries powers safety-critical systems, so it has to be delivered consistently and reliably.

---

70 Not just an example of the age of the technology, but of the kind of backers these technologies enjoy. Richard Feynman was an extraordinary physicist who worked on the Manhattan Project, won a Nobel Prize for work on quantum electrodynamics and contributed to the investigation into the Challenger space shuttle explosion. He also wrote what may be the most popular physics book ever.

Every time we convert energy, we lose some to heat or noise. So storing energy without further transformation is more efficient. Some businesses build air-conditioning units that use the excess (cheaper) electricity produced at night to freeze water, which cools air the next day. Or gravel, molten salt or argon gas store heat before heating homes and offices. As you will see in a minute, excess electricity can produce hydrogen or biofuels, which give us an energy store.

Baseload

Now is the time to mention the ogre of 'base load'. Proponents of fossil fuels and nuclear power say we must have a high, constant supply of electricity (base load), and only their chosen fuels can provide that consistency because renewables are 'intermittent'. Let's break that argument down.

'Intermittent' is a poor description because it suggests that renewables stop producing electricity with no notice. A better term is 'variable'. Considering your home, if your variable solar produces electricity only on sunny days, then you will turn to the grid on rainy days and at night (in other words, the grid supplies a base load). However, if you invest in energy efficiency, buy storage, and jet-wash your car only when the sun shines; then you will need less, if any, electricity from the grid. By changing your energy use, you have reduced your need for a base load.

Another way of reducing the need for a base load comes from managing generation differently. If the grid has two sets of solar panels set a fair distance apart, the chances of both failing to produce electricity is less (during the day) than with just one set. Two wind farms will generate electricity throughout most nights. More panels and turbines further reduce the chance of not producing electricity. You will find some analysis in the Sources that show that in a single year there were only three days when the wind failed to blow across the whole of Europe. You can read this as a need for base load (as the author did) or for the potential to use a mix of renewable sources and storage to reduce the need for 'always-on' generators like coal-fired power stations massively (they are always on because they cannot fire up quickly).

Finally, should all forms of variable renewable energy fail, and stored energy run low, then the smart grid can turn to geothermal, concentrated solar with storage, hydroelectricity or biomass, which fire up quickly. So, with the right investments, we have no need for a fossil fuel base load, just flexibility in the way we use and produce electricity. Look in the Sources to find a book on this subject, along with a mathematical model.

Manage all energy, not just electricity

Energy storage works more efficiently if we challenge the current situation in which heating (natural gas and heating oil), transport (the liquid fuels) and electricity are handled independently. Managing all three in an energy pool gives us more room for manoeuvre. When we are lying in bed listening to the wind howl around the eaves, wind turbines are generating electricity few people need at that time. In a combined energy market, the excess electricity would break down water into oxygen and hydrogen or create biofuels. Hydrogen and biofuels power transport and provide heat, so producing them creates an energy store.

Taken the other way, supply can be boosted temporarily by the batteries in electric vehicles. For example, immediately after the evening rush hour when we plug in our cars, they could power televisions, ovens and kettles. When we go to bed, the cars would re-charge.

Manage demand and supply (load shifting)

The term 'load shifting' might sound intimidating. But all it means is smoothing out demand. Today we mainly increase generation when we need more electricity. In the future, when supply cannot keep up, we will shift much more demand. If we give energy companies control over some of our appliances, they can time our use of them, increasing demand when electricity is plentiful and reducing consumption during peak times. The more engineers work on this subject, the more they find opportunities for managing demand – from domestic appliances right through to industrial equipment.

Of course, you might ask why the utility companies do not use electricity prices as an incentive for us to move demand ourselves. They have two reasons. First, we are not good at managing demand, and would need equipment to do it for us in any case. Second, if all consumers use independent equipment to move to a cheaper period, peak demand would only move, rather than spread. Energy suppliers can smooth demand, and incentives encourage people to hand over control. In most countries, large users of electricity have contracts that provide them with power at a lower price, if the user agrees to reduce consumption when asked.[71] In the United States, energy companies offer controllers that turn on domestic air conditioning to avoid peaks of demand, eg, when people return from work while making sure customers are comfortable and have the smallest possible energy bills. Consumers who sign up for these schemes benefit from reduced tariffs. Trials in Australia and Canada aim to develop equivalent systems. Researchers in the UK have conducted trials focusing on washing machines and air source heat pumps and found that we Brits are keen to give it a go.

Of course, the ideas we saw when we thought about 'managing all energy' give us even more scope and flexibility in matching demand to supply.

Cross borders

Traditional electricity grids follow national and political boundaries. Finding France's nuclear power stations on the boundaries between *départements* (counties) is amusing. However, you are more likely to find highly-prized and -protected electricity generation in the heart of countries. Although this made some sense during the Cold War, isolating supply exposes us to variation in demand and makes variable power supply harder to use. Smart grids help us share electricity.

Our National Grid is working closely with its equivalent in Norway to develop cable and control technologies that will let us buy Norwegian electricity. Power links already join several countries, and more countries are planning to strengthen energy security by sharing with neighbours. Germany, Norway, Sweden and Denmark already routinely share electricity, and as Norway holds more than 40% of the hydro storage capacity in Europe, it could supply storage for the generators in the other countries (eg from Germany's growing renewable sources). Other Scandinavian and nearby countries are also connected – for example, Poland, Sweden, Lithuania, Estonia, Latvia and Finland. Europe has been connected to Africa by cables since 1997; and with investment in

---

71 In the UK, the deals are called reserve services and are managed by National Grid.

EU grids, the expansion of this 'Medgrid' will allow Europe to get more electricity from the Sahara.[72]

Energy efficiency

A smart grid encourages energy efficiency in a few ways. For us average Joes, smart meters tell us how we use our electricity. Armed with that information we can identify appliances which use too much. Friends of the author used a multimeter[73] to work out their fridge was using too much electricity and cut £100 a year from their electricity bill when they replaced it. You are probably not that geeky – in which case a smart meter is ideal, as it will help you do the same thing.

Imagine you have a big open space – a factory or warehouse – and you need lighting. A smart meter helps you understand which lights use the least energy, and to replace faulty units. You can get a comparison with similar buildings and share top energy-saving tips. In a few seconds we will look at the business models of utility companies and return to your warehouse.

Finally, better data about consumption helps the energy industry reduce losses. And they can measure power cuts, identify their causes and prevent reoccurrence.

Charge for managing energy, not producing electricity

A business model is the method of making money a company follows. Utility companies have a simple business model: buy energy from suppliers (gas) and generators (electricity) and sell it on at a profit. The model starts to fall apart with distributed generation and energy efficiency. Good utility companies are already seeking new ways of keeping us warm and entertained while still turning a buck. You might think this will never happen as the media often portrays all energy companies as money-grubbing, head-in-the-sand businesses. But in 2013 PricewaterhouseCoopers (PwC) surveyed senior energy company executives from 53 companies in 35 countries, and found that the vast majority expect significant changes to the energy business model.

---

72 The Sahara is one of the three best places on Earth for solar power. It is part of an area stretching from Algeria to the Middle East. The other two are in southwestern America and western Australia.

73 A nifty piece of kit that can measure voltage, current and resistance.

Solar installers have lit the path to a new business model. One of the changes is selling energy as a service. Say you cannot afford to install solar power, and you do not want to (or cannot) get a loan to finance a deal. Some solar businesses will install solar power on your roof for no fee and charge you for the electricity at a reduced rate, effectively paying you rent for your roof space and a bit of your attic.

Let's go a step further, and return to your big warehouse. You could pay a fixed fee to light your factory, and the energy company would select the most efficient lighting to suit your needs. They have two financial motives to reduce your electricity use. First, direct profit; second, energy efficiency costs less than new generation. There are so many ideas in this area that we can expect each company to develop ways of working to suit their market and ambitions. Unfortunately, we are seeing fewer changes to business models at home than overseas. For example, the UK government runs an activity to help householders improve efficiency and change to low-carbon energy sources, but the scheme was slow to take off, not least because the 'Big Six' are reluctant to change their business models. However, smaller utility companies are pushing for considerable changes to the way we generate and sell electricity in the UK.

Recognise silver buckshot

We keep using the phrase 'silver buckshot'. OK. It means we need to stop looking for a silver bullet: a single way of replacing fossil fuels. Instead, we must develop a good portfolio of alternatives to ensure a continued electricity supply. Our 'buckshot' consists of alternative energy production, distribution, storage and management options.

No alternative to oil works best in all situations, and few can work alone. If you live near a fast-flowing stream, you might invest in a turbine, but your dry-gardened neighbour would find solar more effective. You would both want a connection to the grid; one for dry, hot weather, and the other for nights and rainy days – or perhaps you could pool your resources and buy some storage.

The smart grid is here

The first computer looked nothing like the computer in your phone. Similarly, early takes on the smart grid will not satisfy all our needs; for example, we might need fossil-fuel backup. However, they are a step in the right direction, and they support investment into generation and control technologies. Unless we expect fully-fledged smart grids to appear overnight, we can't ask for a better starting point. When the power plant burnt down on the Caribbean island of Bonaire, the local government made the decision to turn completely to renewables. Efforts started with an integrated diesel and wind power grid, with battery storage. The price of electricity dropped, and attention has now turned to the production of biodiesel from salt-water algae. Even without biofuels, this system pushed the bounds of engineering and is a boon for the inhabitants of the small island.

At the other end of the spectrum, Italy boasts the world's first and largest smart grid, though at this point it mainly consists of smart meters. It makes savings of 500 million Euro per year on an investment of 2.1 billion Euro. The Enel smart grid manages the electricity supply to more than 30 million domestic and industrial users. UK homes and business will need 53 million smart meters. The programme to install these meters will cost £10.9bn and create benefits of £17.2bn by 2030.[74]

The UK's energy watchdog Ofgem looks out for the needs of consumers, generators, suppliers and grid operators, and it supports the development of smart grids and renewable energy. Ofgem aims to improve the health of our energy market; it describes its goals and plans as RIIO (Revenue = Incentives + Innovation + Outputs). You'll find a link to the description of this framework in the Sources. You will also see a letter to the *Financial Times* from the ex-head of Ofgem. In it he says:

> *"The UK model is now being actively discussed in New York and California and is being reviewed across the globe. Exporting regulation – you could not make it up – but it is true!"*
>
> *– Alistair Buchanan, Jan 2015*

Smart grids make sure we have electricity when we need it. Without a smart grid, alternative means of production cannot meet the standards of energy supply we take for granted today. If we want to get past fossil fuels, we must continue to invest in smart grids. The rest of our investment will fund generation.

---

74 Those numbers are right; the Italian system paid for itself in four years, whereas we need ten or so. If you are like me, you will want to know how the Italians did it for so much less money and to emulate their example.

Smart grids do have a disadvantage: they have introduced the unfortunate word 'prosumer' to the lexicon. The term recognises that people and businesses will both provide and consume energy, so it is a step forward, if a little ugly.

Electricity generation

We are all, naturally, cynical about new technologies; however, in a moment we will see that the major renewable-energy technologies are far from new. Before that, let's see what the people who are paid to provide financial and energy advice to businesses, investors and governments are saying about renewable energy. In June 2015 analysts at UBS, an investment bank, declared that solar will, eventually, supplant nuclear and coal. And in March 2015 the University of Cambridge and PricewaterhouseCoopers (PwC) published a report for the National Bank of Abu Dhabi. In it they concluded renewable energy technologies are capable, ready and cheap enough to replace fossil fuels.

If you have come into this chapter believing that renewable energy and other low-carbon sources such as nuclear can never replace fossil fuels, remember those findings as you look at the following dates.

Solar – heating space and water since forever. The photovoltaic effect was discovered in **1839** and first used commercially in 1955.

Wind turbines – first proven in **1887**.

Tidal power – used by the Romans and Chinese, popular in Europe in the Middle Ages. Producing electricity since **1966**.

Hydroelectricity – the energy of choice in **1881**. Closely followed by pumped hydro, first used in 1890.

Ocean Thermal Energy Conversion – first proposed in the **1880s** and producing electricity off the coast of Brazil in 1935.

Nuclear fission – first used in electricity production at Obninsk, Russia in **1954**.

Geothermal energy – used by humanity for heating for millennia. First produced electricity in Tuscany in **1904**.

Burning stuff – first used by Edison on Pearl Street in **1882**.

Energy efficiency – popular since the **Seventies** and growing as a form of 'capacity creation'.[75]

These technologies, which individuals, communities and businesses have used for decades, will release us from our bonds of oil. Our job is using and developing these new technologies while we can still afford to. Use and development drive down costs, improve performance, and let us understand and resolve risks. While this chapter focuses on technology, in "To oil, or not to oil?" we reflect on the vexed subject of government support, one of the factors that influence how well we develop new technologies.

Solar

French scientist, Edmond Becquerel, first discovered the photovoltaic effect in 1839. Generations of scientists and engineers have studied it; Einstein won a Nobel Prize for his work explaining it. In the simplest of terms, sunlight causes electrons in semiconductors to move and create an electrical current. Since the early days, the potential of photovoltaic materials has soared. We can build solar cells into windows and incorporate them into fabrics.

Solar is a significant contributor to overall electricity generation. For example, by July 2014 the UK had 3.8GW of solar photovoltaic capacity. Some of that capacity is in solar farms, but most on rooftops (2.7GW at the end of 2014), and businesses looking to reduce energy costs use the technology in their factories and offices. Dow Jones (the money people) completed a 4.1MW installation in their New Jersey offices in June 2011. They claim it can power 1,230 average homes.[76] They are far from being alone; you will find oil companies installing solar power to create steam to loosen oil reserves in Saudi Arabia and California, to name just two locations. Remembering the oil expense of mining, it is good to learn that

---

75 Buzzword bingo anyone? This phrase means if you stop using the electricity from a power station, you effectively create more capacity to supply the remaining demand.

76 The 'average home' is a measure intended to help us all understand the size of 'power stations' without worrying about gigawatts and megawatt hours. However, it is not a standard measure. To get an idea of the size of this installation in terms of UK homes, we can use Department of Energy and Climate Change data, which shows the average UK home uses 16.28MWh of energy per year (including transport fuel). Assuming the Dow Jones facility operates at 60% efficiency for half the time, it will produce enough electricity to offset the energy used in 661 UK homes.

a potash mine in New Mexico and a copper mine in Western Australia will use the power of the Sun's rays to reduce their energy bills. Closer to home, Blackfriars Bridge in London uses solar panels to provide electricity to Blackfriars station.

Sticking with sunshine, we can use it directly. Heating water with solar is growing more popular in the UK and 'passive houses' absorb sunlight to heat rooms. The UK government reports that in 2013, UK active solar produced 414GWh of energy.

A Concentrating Solar Thermal Power (CSP) plant would not fit on your rooftop. These behemoths use mirrors to gather and concentrate sunlight and turn water into steam to power turbines. They have some disadvantages. They are harder to build and run than their photovoltaic cousins. And because the plants are big, we have few suitable locations, and they kill animals from tortoises to birds. But, if we can halt their ecological impact, they have a big advantage; they can store heat and provide electricity after the Sun has gone down. Companies build dual-fuel systems to keep the turbines turning during prolonged periods of insufficient sunlight. Backup fuels can be fossil, one of the other items we will think about burning later, or even geothermal.

Another use of solar power is cooling air and water. We already know the Universal Exposition in Paris in 1878 demonstrated the technique. If we look to southern Austria, we can see history repeat itself at a winery that uses only water heated by the Sun for cooling and power.

We will end our round-up of solar technology by hitting the road. The approach taken by the Worcester Polytechnic Institute, Massachusetts uses a 'normal' road, which is built with pipes. Water in the pipes draws out heat, which is used to produce electricity. As an added advantage, the cooled road lasts longer, so we save oil in two ways. At the other end of the spectrum, American entrepreneurs hit the headlines in 2014 with photovoltaic panels that can be assembled to create a road surface or car park.

Solar has been the energy of the future for so long that you might be cynical about the experts who say the price of electricity from solar power matches that from fossil fuel-fired power stations. Price parity, its Sunday name, has reached sunnier climes and could arrive in the UK before long.[77, 78] Price parity has a better pedigree (and is a more modest goal) than Thomas Edison's old, bold boast that "We will make electricity so cheap that only the rich will burn candles."

Wind turbines

Summer temperatures in the Antarctic interior are rarely higher than -20°C, and in winter it experiences the coldest temperatures on Earth.[79] Yet this harsh environment is home to communities of scientists, who live year-round on the frozen wastes and need energy to survive and work. Powering life in this remote and barren land brings transport challenges because ships can deliver diesel only when the weather permits. Moreover, avoiding pollution is a point of principle, so renewables are a no-brainer. At the bottom of the Earth, although photovoltaic cells are super-efficient, they only work for half the year.[80]

Luckily, Antarctica also experiences some of the highest winds on Earth; speeds of up to 60mph are common.[81] Turbines on the Mawson Station (Australian, and the oldest continuously-inhabited station south of the Antarctic Circle) harvest the energy in wind and satisfy 95% of the station's energy needs most of the time. Other countries are also investing in Antarctic wind power. These systems help develop control methods and building techniques that will be useful in friendlier settings.

Back on the more highly populated continents, the Chinese government intends to move toward decarbonisation, by which they mean reduce fossil fuel use. Carrying out the policy is resulting in massive building projects and has increased China's installed wind energy capacity from 146MW to 114,763MW (114GW) in the period 1997 to 2014. By

---

77 It has already reached the Republic of Ireland.

78 Scientists and engineers will give us price parity in two ways: the good old-fashioned learning curve, and new technologies. We can help reach price parity by buying solar and supporting the solar industry.

79 The mercury fell to -89.2°C during the winter of 1983 at the Vostok Station, a Russian station 1,260km from the nearest coast.

80 It only works for half the year everywhere on Earth, but the division between day and night is more extreme at the poles.

81 Surprisingly, these speeds are not the fastest measured. Mount Washington in New Hampshire holds the record for the fastest ground-level wind recorded with an anemometer outside a tropical cyclone: 231mph.

2020 China will have 200GW of installed wind capacity, compared with the UK's total electricity generation capacity of 85GW (in 2014).

While the world's largest wind farms are in China, wind farms are a solid contributor to power generation everywhere. As of February 2015, Europe had an installed capacity of 128,800MW or around 14% of installed capacity. In a typically windy year, these turbines will produce 284TWh (terawatt hours) of electricity: approximately 10% of European demand. From 2000 to 2014, 29% of all new European capacity was wind.

But the claims of opponents give this form of electricity production a strong headwind. They complain about the look of turbines, noise and other health concerns, increasing electricity prices and plummeting property prices. They give wind turbines the emotive nickname 'bird chompers'.

Aesthetics are important and personal. A popular viewing point on Spain's south coast looks over magnificent cliffs into the Mediterranean. Yet virtually every camera looks inland at the huge wind farm marching along the coast; clearly not everyone dislikes turbines. The wind farm consists of turbines of various designs. If you wanted to create a museum of wind turbine technology, this would be a good place to come (it's kind of cool even if you don't fancy curating an exhibition).

Besides, not all wind generation relies on a set of turbine blades on columns. Thinking laterally, why put the turbine on a pole? We can install vertical columns that resemble garden ornaments, albeit larger. The Bahrain World Trade Building has two sections funnelling wind into the space between them, where it turns wind turbines hanging on cross-spans. The turbines produce 18% of the electricity used by the occupants of the towers.

Concerns about the way wind turbines look and the variability of low-altitude winds have inspired several groups of engineers. They are chasing high-altitude solutions. Rather than working alone, these innovators have come together under the Airborne Wind Energy Consortium to share ideas and resolve problems. Each group is exploring a different way of producing electricity from the almost constant winds that blow between 200 and 5,000 metres above the ground (aeroplanes fly at 10,000 metres). The technologies range from helium-filled blimps to wing-like structures and kites; each technology has advantages and challenges, and the ability to become part of the non-fossil fuel electricity mix. Indeed, the citizens of Fairbanks, Alaska will soon claim the world's first airborne turbine.

However, ground-based turbines are still the most common form of wind-powered generation and turbines can be noisy, so planners site large-scale wind farms away from populated areas. There are two elements to the noise: infrasound, and amplitude modulation - sometimes called swish and thump. A UK government-sponsored report found no evidence to back up claims of health damage from infrasound. So what about that swish and thump? Well, scientists and engineers are developing turbines to be less noisy. Not only will this make wind power more acceptable, but it will also let blades turn faster and generate more electricity. You might be sceptical and believe that a piece of metal passing through the air must always be noisy. So imagine the soundless flight of an owl (just don't imagine being the mouse at the end of that journey). Scientists at Cambridge University did just that and have demonstrated that a 'sound scattering' material can reduce the noise of turbine blades without compromising aerodynamics.

Many people observe that communities who buy wind turbines do not complain of noise or ill-health or the look of wind farms while those with wind turbines forced on them do. A group of psychologists has taken that finding a step further. In a carefully constructed scientific experiment, they found people with negative expectations felt unhealthy around turbines. Confirming noise and sleep are not a problem to most neighbours of wind farms, studies in the UK and United States disprove claims of hugely reduced house prices.

Of course, some turbines are unreasonably noisy and their owners must make repairs.

In April 2010 an independent study conducted for the European Wind Energy Association showed the overall cost of electricity decreases if wind power contributes to the mix. As you might expect, the minute-to-minute price varies with the fall and rise of the wind.

Finally, the bird (and bat) chomping. Any structure that stands in the flight path of birds will result in bird deaths. In particular, some early, notable wind farms, such as the Altamont Pass in California, were

> *"sited with very little consideration for the indigenous raptor populations".*

> *– Centre for Sustainable Energy*

Now, planning for all wind farms considers the safety of birds (though this may be more true in Europe and the US than in other countries). Scientific data taken from the United States and Europe in 2005 showed wind turbines caused one in 10,000 premature bird deaths.[82] The total equals the number killed by aeroplanes and is 1/1,000 of the corpses created by cats. Buildings kill more than half the birds that die before their time. A more recent study shows American turbines kill 2% of the Golden Eagles wiped out by humanity. But the industry is not resting on its laurels; it is testing new designs, better siting and changing patterns of operation to reduce the number of bird deaths. Golden Eagle experts in the United States are supporting development work by monitoring the behaviour of Golden Eagles. In the UK, the RSPB reviews hundreds of planning applications each year and objects to those that threaten bird populations.

As birds are killed in many locations, such as airfields and around factories, wind farms could benefit from best practice in other industries. Clearly no one near a wind farm would appreciate the use of explosives to scare off avian visitors so we may turn to the use of lasers, which birds find very, very frightening.

According to the United States Geological Survey, no one knows why wind turbines kill bats, but they do – and in large numbers. Because we know little about bats, no one expected their deaths. Indeed, scientists thought bats would detect and avoid the spinning blades. While some species might, it seems the pressure changes around turbines kill some bats instantly. And others may find the turbines attractive during migrations and mating seasons. The British and American Bat Conservation organisations both support wind farms. They work with planners to ensure protection for bats and undertake independent research. Changing how we operate wind turbines and setting up deterrents will reduce bat mortality.

So far we have stayed high and dry, so let's dive into offshore wind farms. Putting wind turbines at sea resolves several issues for wind power and introduces new ones. Bringing the electricity to shore might be a problem, but, as we saw a few pages ago, undersea cables have a good pedigree. The increasing depth of the seabed poses another problem. It

---

82 Some people dispute the validity of such studies, claiming that wind farm operators purposefully miscount bird deaths. To ensure accuracy of the count, specially trained dogs sniff out bird (and bat) carcasses.

is quickly too deep for turbines. We will encounter the same issue in "What has the oil industry ever done for us?" and the same solution. Both industries use floating platforms. Buoyant turbines have two benefits: they move power production farther from shore and concerned citizens, and they open more wind for exploitation. The UK's first floating wind farm will be commissioned in 2017. It is a pilot to demonstrate the concept. Of course, you remember the kites in "Here, there and everywhere"; they provide a potential alternative to turbines.

Tidal power

Tides hold much energy, and many ideas for capturing that energy are in development and use. The world's largest, and oldest, tidal power station is in Rance, France. It has a 240MW capacity, about the same as the UK's hydro stations combined. Tidal power is not new on the Rance river. Here tidal mills turned until World War II and the first experiments in electricity generation took place in 1921. Opened in 1966 after five years of construction, the Rance hydropower station lies in a large dam, 330 metres long, across the estuary of the river. The dam, or barrage, allows water to enter the river basin during high tides. As the tide turns, the barrage closes and forces the water through turbines to produce electricity.

The UK government has decided against funding a similar barrage across the River Severn. So, unless private funding can be found, [83] we have to look for smaller tidal schemes to see progress in our waters. For example, two large turbines, attached to a column fixed to the seabed off Belfast, produce electricity for the city. The assembly weighs more than 1,000 tonnes and has a 1.2MW capacity, enough to power more than 1,000 average homes.[84] And similar structures will soon be found in the Pentland Firth in the world's largest tidal array.

Staying out at sea, we could soon find tidal lagoons. Though they work in the same way as barrages, lagoons sit on the coast and have less

---

83 If you are asking why the government should provide financial support, hold your horses until you get to "To oil, or not to oil?".

84 Just like the Dow Jones example, these figures do not match Department of Energy and Climate Change data, which show the average UK home uses 16.28MWh of energy a year (including transport fuel). Assuming the Belfast facility operates at 90% efficiency for three-quarters of the time, it will produce enough electricity to offset the energy used in 435 UK homes.

environmental impact. Developers have planning permission for the world's first tidal lagoon in Swansea Bay (320MW). However, at the time of writing, they are waiting for a decision from the Department of Energy and Climate Change (DECC) for funding. The same company would like to build a further five lagoons across Wales and England, which it says will supply 8% of the UK's electricity requirements and promote the UK as a source of innovative low-oil energy technologies. You may have noticed that these lagoons are similar to the technology proposed for energy storage in Belgian waters.

Hydroelectricity

Back upstream, rivers have powered mills for centuries; hydroelectricity is a favoured descendent of that technology. Today we harness the power of water in three ways: dammed rivers (also called conventional hydro), run of the river, and pumped hydro. Dammed river systems drive massive turbines and power the world's five largest power stations (the sixth is nuclear), and worldwide we have 190 hydroelectric plants bigger than 1,000MW in operation or under construction.[85] Run of the river hydroelectric schemes are smaller than their conventional cousins and have much less water storage (think small lake, rather than reservoir) or none at all. In British Columbia, Canada more than 700 individuals, businesses, consortia and communities have installed run of the river systems, and we will meet another in the section about people who use renewable energy.

As we know, pumped hydro stores the energy in electricity. In the UK it can store a little over 3% of our electricity and release it at peak demand. We have four pumped hydro stations in the UK: two in Wales and two in Scotland. Italy and Spain were the first countries to use pumped hydro commercially, in the 1890s. However, large-scale pumped hydro came in the 1930s, with a turbine that could generate electricity and pump water. Since then, schemes have grown much larger. Now

---

85 By size, two of the top four dams are in China: Three Gorges Dam 22,500MW and Xiluodu Dam 14,000MW. The others are the Itaipu Dam, 14,000MW, on the border of Brazil and Paraguay (it supplies both countries) and the Guri Dam in Venezuela, 10,200MW. Ethiopia plans to build two dams and courts controversy. The dams will have a huge effect on the area that will become a reservoir and will change downstream flow patterns and rates. The Congolese government is planning the world's largest dam on the Congo River (Grand Inga).

"large" is 1,000MW or greater. Before you get all sizeist, 1,000MW is a big power station. In 2013 the UK had more than 400 power stations, only 28 of which were 1,000MW or greater.

Not all moving water is tidal or flowing

In the introduction to this section we learned that the wind creates waves and waves have energy. We harvest wave energy in many ways. At the Limpet power station on Islay, rising and falling water levels force air through a turbine. In other systems, equipment on the surface of the water will use the energy of the wave to drive a turbine with water or hydraulic fluid, and off the shore of Perth, Australia, enormous buoyant discs will push and pull on a seabed actuator. Some systems will generate the energy offshore; others plan to pipe the fluid back to land, to a generating station (potentially providing storage).

The UK is investing in wave (and tidal) energy and has set up three centres to develop commercial systems. They are the European Marine Energy Centre (EMEC) in Orkney, the Offshore Renewable Energy Catapult (formerly NAREC) in Northumberland, and Wave Hub in Cornwall. The centres provide test resources and consultancy to businesses developing tidal and wave energy. They develop marine energy standards with other international bodies. Let's hope at least one of them will test our next sea-based technology.

Currents below the waves move continuously – so you might wonder whether the energy in ocean currents could turn turbines. At high speeds – say, at the Florida end of the Gulf Stream, which boasts four knots – they can, and people are working on that question, but at depths of 300 metres they face substantial technological challenges. However, low-speed water cannot turn turbines, so Michael Bernitsas of the University of Michigan approached the opportunity from a different angle. He developed a way of producing electricity called VIVACE, which transforms vibrations into electricity. Just looking at the maths, the technology need not be very efficient as just 0.1% of the ocean's energy would satisfy the demands of 15 billion people. Bernitsas does not claim he can channel even that tiny proportion; still, his technology is exciting.

Ocean Thermal Energy Conversion

Since 1880, engineers – including Nikola Tesla – have tried to generate electricity using the differences in temperature in the sea. Tesla later declared we would never produce large amounts of electricity this way. However, engineers and scientists persevered and proved the technology in the 1930s (off Cuba and Brazil). Since then, the Japanese government has funded work that resulted in the first commercial plant, installed in the Republic of Nauru in 1981 with a capacity of 120kW. In 1999 the United States took the technological lead and built a 250kW generator.

Here comes the science. The seas are cold, with warm water only at the surface – the difference in temperature between the surface and deeper water can be as much as 24°C. Ocean Thermal Energy Conversion takes advantage of this contrast to generate electricity.[86] It also has other benefits, like desalinating water, extracting nutrients and minerals from the sea, and producing hydrogen. It can provide cold water for chilled-soil agriculture, which helps plants from temperate zones grow in warmer, dryer subtropical regions.

If it's so good, why has no one built another plant? Well, fluctuations in the price of oil mean that investment in alternatives is sometimes high on the agenda (for example, in the Seventies) and sometimes low. The volatile, fickle, unpredictable, undependable price of oil, along with climate change, has blown the dust off this technology. Hawaii now has a 100kW system, and Lockheed Martin is starting the detailed design and manufacture of a 10MW demonstration plant, which they will build off the south coast of China.

Nuclear fission

Back on dry land, we find controlled nuclear reactions producing heat (which creates steam to turn a turbine). Nuclear energy first produced grid-connected electricity in 1954 in Russia.[87] Then the oil shocks of the Seventies motivated France and Japan to reduce dependency on oil for electrical generation – to the point where, before the 2011 Japanese

---

86 We have two main methods of using the nearness of warm and cold water to produce electricity. In closed systems, warm water heats a liquid (eg ammonia), which evaporates and turns a turbine. Cold water condenses the ammonia for reuse. In open systems, warm seawater evaporates, producing low-pressure steam which turns a turbine. Cold water condenses the steam, which can be returned to the sea or pumped elsewhere.

87 Yes, it beat Calder Hall (at Windscale) by two years. Though Windscale can claim the first fire in a nuclear plant (in a plutonium breeding reactor, not the power station itself).

earthquake and tsunami, they enjoyed the highest levels of nuclear power after the United States.

Public opinion on nuclear power wavers. After the accidents at Three Mile Island (1979) and Chernobyl (1986) we backed away from it. In the face of rising fossil fuel prices, countries ranging from China and the United States to South Korea and South Africa put nuclear generation back into their energy plans. When the 2011 Japanese earthquake and tsunami led to failures and leaks at the Fukushima power plant, people and countries decided they do not want to use nuclear fission for power. However, construction of Fukushima Daiichi started in 1967 and ended in 1971. Slightly up the coast, the Onagawa plant, built in 1980-1984, suffered a fire, but no leaks of radiation. In fact, 200 people sheltered there after they lost their homes in the tsunami. Further, Fukushima's designers only reckoned for earthquakes with magnitudes up to 8.6; the tremor that destroyed its backup generator was four times stronger at magnitude 9.0. And studies and data from the pro-nuclear lobby tell us Fukushima only released 1/10,000 of the radiation released by fossil fuels in a year. For a colourful comparison, the 165 million bananas we eat every hour of every day contain twice as much radiation.

Of course, there is scope for reducing the risk of accidents. One way is to reverse the trend to larger stations and set up small-scale reactors. A few US companies are working through the implications – regulatory and financial – of smaller reactors. Their basic premise is that small plants can be simple, and that simplicity removes the need for mechanical and electronic controls (which can go wrong) because gravity and other physical characteristics will kick in if the process becomes dangerous.

Changing radioactive material also does away with the need for complexity (though not to the same extent). If we use thorium instead of uranium, we can use salt instead of water to absorb heat from the reactor and apply it to generation. Changing to salt reduces the pressure of the system, the risk of containment failure and the complexity of control systems. And thorium can replace uranium in many existing reactors, which could be a cheap and quick way of adopting this less hazardous material.

Thorium is a good fuel because it does not hold enough radioactivity to create a chain reaction. That might seem backwards as thorium needs an energy source (usually neutrons) to decay. But it means operators can quickly switch off a thorium reactor. Thorium has other advantages over uranium; for example, it cannot fuel bombs, and it produces fewer long-lived elements as it decays. It also provides much more energy; you would need 250 tons of uranium to produce the same energy as one ton of thorium.[88]

After the fear of disasters and weaponisation, the biggest problem with nuclear fission is waste storage – though "storage" sounds a bit temporary when we are talking about thousands of years. In virtually every country with plans to create deep storage for nuclear waste, the plans are on hold. Understandably, no one wants nuclear waste close to them, particularly when we cannot hope to predict the geological changes that will occur in the future. On the other hand, the nuclear industry reminds us that they alone take responsibility for power generation waste and include the cost of safe disposal in electricity prices. Here are some numbers:

- Each 1,000MW facility produces between 275 and 425 cubic metres (m^3) of radioactive waste per year (including the glass used in vitrification). Each coal-powered station of the same capacity produces 400,000 tonnes of ash, airborne pollutants, such as sulphur and nitrogen compounds, and much carbon dioxide.

- When considering toxic waste, we in the OECD countries produce three million tonnes, of which conditioned radioactive waste is about 81,000 m^3.[89]

- In the UK 94% of conditioned radioactive waste is low-level, around 6% is medium-level and 0.1% is high-level. However, that tiny proportion of high-level waste contains 95% of the radioactivity.[90]

Still, that high-level waste is troublesome. So, German scientists are developing a form of nuclear power generation that uses liquid instead

---

88 While we have focused on thorium here, in common with biofuels, the nuclear industry talks about generations of development. For instance, an international collective is testing a range of 'generation iv' methods. These methods are much more advanced than our current 'generation ii' plants; many can use materials previously classed as radioactive waste.

89 That isn't a good comparison as it mixes units. If we assume that conditioned radioactive waste has twice the density of water, ie each litre weighs 2kg, then we can calculate the weight of those 80,000m^3 to be 160,000 tonnes, or 5% of all toxic waste.

90 Interestingly, the tailings from uranium mines are not, 'strictly speaking', classed as radioactive waste.

of fuel rods. Liquid fuel has two advantages. The fissile material in spent fuel is easier to recycle, and the remaining liquid has a radioactive life of around 600 years, compared to thousands of years for solid fuel waste. And so-called fast reactors, which turn radioactive waste into fuel are gaining popularity, though also still only on paper.

Did you read the sidebar explaining 'Why non-fossil fuels are not yet ready to be our sole source of electricity'? It discusses the technological and political reasons why we are not ready to use renewables as our only source of energy. While nuclear fission has the benefits of high efficiency and energy density, it suffers from adverse economic and political factors. New technologies must pass stringent safety tests; currently, these tests are based on large-scale, water-cooled uranium reactors. If we are to move to a new generation of reactors, we have to be willing to invest in developing appropriate safety standards and to spend the money required to demonstrate compliance. Luckily, countries and businesses are taking on this burden.

Thor Energy (an international consortium) started the production of energy from thorium in a Norwegian experimental facility in July 2013, and they plan to have a commercial plant in operation in 2020. China and India are also each building trial thorium plants, and dozens of new companies are looking at a wide variety of technologies to build small scale reactors. Our job is to be open-minded to the potential of nuclear power while creating and maintaining proper safety standards.

Geothermal energy

If we turn to the Earth for heat, we avoid direct use of radioactive isotopes and the opposition they bring. Iceland is a sterling example of making the most of the resources to hand (or under foot); this rugged country produces almost 100% of electricity (mainly hydro) and 85% of primary energy use (largely geothermal) from renewable sources. Renewable heat is a benefit of living on an island created by volcanic activity. Even though the UK is much less geologically active, the ground below our feet holds heat travelling from the Earth's core, and we could make much more of it.

Hot Dry Rocks is not just a great name for a band. Several kilometres below the surface, the Earth's crust consists of igneous and metamorphic rocks – for example, granite. The rocks are (all together now) hot and dry. We can use them to heat water and use the hot water to turn turbines (to produce electricity). Cornwall has the right conditions for geothermal energy, and the Duke of Cornwall (Prince Charles) supports the method. Another supporter of deep geothermal is the City Council of Stoke-on-Trent, which is planning to build a district heat system to provide heating to homes and businesses along an 11km pipeline. Stoke can do this because it lies on top of an ancient volcano. In some geothermal plants, fracking[91] makes a path for the water. Just like the earthquakes, we will discuss in "Where has all the oil gone?", pumping water into fractured rocks has caused earthquakes in Switzerland[92] and California.

Hot Dry Rocks is a modern form of geothermal energy, which we can exploit in many areas because it does not rely on co-located water. Older systems, including the world's largest at The Geysers in California, take advantage of water already present in the area.

Clearly we can't all drill down as far as geothermal plants do (over 4km). Fortunately, the first 100 metres of the Earth (give or take) is as warm as the mean air temperature. So, Siberia has permafrost, and British temperatures sit between eight and 11°C. Ground-source heat pumps transfer the heat to air or water. Systems cover a wide area, with pipes a few metres below the surface, or drill holes to the depth of 100 metres. Switzerland, a conservative country, has more than 70,000 installations. Strictly speaking, this energy is not geothermal. It is solar.

Burning stuff

We mustn't lose sight of our goal; we can still burn stuff, and, if we focus on sustainable sources and methane, we can reduce oil consumption and address some of the concerns of climate change.

The depths of landfill hold little oxygen, so rotting food and other carbon-rich items produce methane instead of carbon dioxide. Some landfill operators treat the gas and sell it into the gas grid, to utility companies and businesses. Because methane heats the atmosphere faster than carbon dioxide, this is a great way of limiting global warming.

Ironically, we can also force processes that usually produce CO_2 to create methane. If we burn the methane, the net impact on the environment is zero, and we can benefit from the energy released. One

---

91 Yes, *that* fracking.

92 One of the Swiss tremors happened when flammable gas threatened a drilling operation. The water was pumped into the well to stop the gas escaping.

way of creating methane is with anaerobic digestion. It works by feeding waste to microbes, in a controlled, accelerated way, and can take any carbon-rich source of waste, from grass clippings and paper to sewage. Indeed, in Mexico, pigs are getting in on the act of producing electricity from a readily available substance.

In the UK in 2014 'thermal renewables' (non-fossil-fuel stuff we burn, such as sewage gas and straw) provided the equivalent energy of 7.3 million tonnes of oil. Wood, the most common form of mass produced 'biomass' contributed 23% of that energy. Unfortunately, when done poorly biomass destroys natural forests, replacing precious ecosystems with only one type of tree, eg eucalyptus.[93] Done well biomass uses energy-rich materials that are wasted otherwise.

Rather than putting biomass into traditional power stations we can burn it close to its source in combined heat and power plants (CHP for short). CHP has two advantages over more traditional power generation: These plants work more efficiently than conventional boilers, and they can use many fuel types. The proportion of renewable fuels used in CHP has increased to over 11%. From heavy industries (oil, chemical, paper and publishing) to light (hotels, leisure centres and universities), CHP disposes of waste and reduces fuel bills. In 2014, UK businesses produced 20.3TWh (terawatt hours) of electricity and 43.3TWh of heat in our 2,066 CHP plants. That works out as 6% of our electricity. Owners of CHP plants sold 9.3TWh of the power they produced (46%) to electricity suppliers and 10.2TWh of heat (24% of that produced) to commercial and domestic buildings (often through district heating schemes).[94]

In the home, we can replace the humble gas- or oil-fired boiler with a micro-CHP unit (and by the end of 2012, 423 of us had). While still providing hot water and central heating, they also produce electricity from energy that would otherwise escape as wasted heat. Although micro CHP units often use fossil fuels, the reduced reliance on centrally produced electricity would result in fewer large fossil-fuel projects. For example, if 15% of UK homes bought a 1kW micro CHP unit, they

---

93 Planting only one type of tree is called a monoculture. Replacing acres of mixed forest in this way has a devastating effect on plant and animal life.

94 Waste is a good source of energy and a better source of materials. Burning it to avoid oil production creates a demand for oil to make new stuff. Such dilemmas mean we have to design our energy mix carefully.

would reduce demand for centralised generation by 640MW, equivalent to one medium-sized, coal-fired power station, like the one in Enfield in London, Coolkeeragh in Northern Ireland or Baglan Bay in Wales. (Scotland gets much of its electricity from renewables and does not have a fossil-fuelled power station of the right size to use as an example.)

One downside of burning is that it produces carbon dioxide and has the potential, when done badly, to create other hazardous forms of pollution. The right technology can overcome both concerns. In "Here, there and everywhere", we met methods of making fuel from carbon dioxide; this gives the CO_2 another opportunity to provide energy and has the advantage of working as an energy store.[95] The other downside is that much of what we waste would provide better value if composted or recycled.

Energy efficiency, again

We have already said that energy efficiency is cheaper than building new capacity, so before spending our reserves (of time and money) on generation, we must review how much energy we waste. The previous chapters covered energy efficiency in industry and transport. That leaves light users like homes, offices, schools and public buildings, and the energy companies themselves.

We can improve the energy efficiency at home, school and work in many ways:

- Reduce heat loss with improved insulation and sealing.
- Turn off appliances when you do not need them, and don't use standby.
- Buy and look after energy-efficient appliances.
- Improve your controls, stop heating the spare bedroom and better manage the times you turn on the heating.
- Convert to DC.
- Use something else.

---

95 Oops, have I already said that? I'm glad you're finding this interesting enough to remember!

Insulation and sealing

There is a good chance you're heating the air outside your home or workplace. To save money, you can take an energy efficiency survey to understand what insulation would work best for you. Even better, you may be able to get financial support from local or national government to help with the cost of insulation.[96]

Turn off

You can find umpteen statements on the Internet about standby and the grand totals made by many tiny amounts of electricity. Although modern appliances use less power on standby, we all still have old equipment. Between us, we have hundreds of millions of pieces of kit plugged into the wall, doing nothing. TV sets are the biggest culprits, but if you look around your house, you will find many more. In your home, wherever a light shines – perhaps on a DVD player, set-top box, computer, monitor or printer – there is standby, and wasted energy. Do you rely on digital landline phones, routers and modems, and signal boosters always being on? Do you leave the charger for your phone, camera or drill plugged in? Do your digital picture frames or the clocks on your oven and microwave display to an empty room? If the answer to any of these questions is yes, then you can improve electricity efficiency without compromising your way of life.

Possibly you spend no more than a few tens of pounds on standby each year, but the power wastage rapidly adds up. The Internet groans with calculations that show levels of waste equal to 'the output of 26 average-size power stations' in the US or 'the energy used by Greece or Portugal'. You will find some analysis for the energy wasted by leaving on computers in "Read oil about it".

Energy-efficient appliances

Over time, appliances become energy-hungry. Dust collecting on the back of fridges reduces efficiency. If you open your oven door because you cannot see through the glass, hot air empties into the kitchen. A dripping hot tap tells your combi boiler to keep boiling. Small acts of

---

96 The dread of cleaning out the loft stops people insulating. So think about the cash in your attic as well as reducing energy bills.

maintenance reduce fuel bills and the demand for oil. When the time comes to buy a new appliance or your smart meter shows your fridge is gobbling energy, be clever and act to reduce your fuel bills when you buy by choosing a replacement with the best possible energy rating.

Controls

Here is a personal example: In our oil-heated home (one of around a million such households in the UK[97]), changing the heating and hot water controller reduced oil use by a third. If you do not have a decent timer and thermostats, you could be heating rooms or water unnecessarily.[98]

Convert to DC

It seems AC may not be the best after all. When the electricity arrives at your home, office, or school it is AC. If you invest in home generation, you will produce DC and convert to AC. When you charge your phone, you convert it back. Each conversion loses energy (turns it to heat or noise). DC powers computers and gadgets. Lighting, fridges, ovens and heaters can run on DC. DC microgrids[99] take AC and DC electricity and deliver it to devices, lighting and domestic appliances. Such grids reduce losses and make battery storage more efficient. Not convinced? You have heard of a DC microgrid, even if you don't know it. Air Force One (the aeroplanes of the President of the United States) enjoy a secure energy supply and gourmet cooking. Their DC microgrid powers gadgets from smartphones to fridges.

---

97 People without access to mains gas are more likely to be in energy poverty than their gassy compatriots, and that is 8m homes in the UK.

98 You might wonder about the perennial debate: whether you would be better leaving the heating on all the time or only turning it on when needed. If your house, office, or warehouse has good insulation, then keeping the heating on avoids the need to raise the temperature much and makes a good strategy (providing you use thermostats). If your building has poor insulation, then keeping the heating on heats the surrounding area as much as the building, so you should restrict heating to the time you need it (you still need thermostats). If your building falls somewhere in between, then you might run some trials to see which works better for you (a smart meter would make this easy to do).

99 A microgrid is a grid inside your building (or aeroplane).

Use something else

Sometimes we turn to electricity too readily. Imagine you have a bank of computers creating oodles of heat. To keep them working, you have to cool them, which uses a lot of energy. You can make it more efficient if you have a nearby heat sink – say, a large body of water. Then you could pump cold water into your buildings to remove the heat and reduce your electricity use by 94%, which makes storage and other forms of generation even more practical. Mauritius is investing in seawater air conditioning to stake its place as a communications hub for the Indian Ocean region. By using cool water, Mauritius follows the lead of Curaçao and Hawaii. We are an island nation, and our water is colder. What are we waiting for?

Efficiency in the energy companies

So much for us; let's discover how the energy companies can improve. Our electricity grid is old, built from the first distributions systems, which developed piecemeal. It was not designed, and is not good enough, to move electricity in the way we need today. In 2014 we lost around 8% of our electricity (28.6TWh) in transmission losses (from power station to substation), distribution losses (substation to consumer) or from theft (997GWh or 3% of all losses). There are three sizeable opportunities for the electricity companies:

Better technology: For example, losses can be reduced by replacing bulky copper wires and parts with the latest superconductors, or adding carbon nanoballs to the insulation of high voltage cables.

Minimise the distance between production and use: For example, in the UK, the National Grid calculates that losses occurring in transmission from the north of England to the south coast run to 12%.[100] New power generation in the south of England is unlikely. Therefore, National Grid and Scottish Power Transmission are investing in high-voltage Direct Current (HVDC) links from Scotland to the west and east coasts of England. 'Wait a minute,' you cry, 'doesn't DC lose loads of power

---

100 I know this doesn't equate to the previous figure; I am assuming that the overall transmission losses are an average and in-region losses are much smaller than the north-to-south losses.

during transmission?' Well, electrical engineers have worked out that HVDC has smaller losses than high-voltage AC. Transformers on land convert DC to AC for further transport (poor old Edison – he was right after all).

Better control and management: As we have already seen (in Leighton Buzzard), better control systems can load-shift, and delay the need for new generation. Smart grids, whatever their size, can improve efficiencies.

The downside to energy efficiency

There is a behavioural problem with energy efficiency. It has the colourful name 'squeezing the balloon', which aptly describes the air (energy) moving from one part of the balloon (energy use) to another. It is also called *rebound*, but that isn't as much fun. Historically, individuals and companies have spent their energy savings on more energy, with the result that energy efficiency measures have led to increased energy consumption.

Lots of very smart people have modelled the possible rebound effect for future energy efficiency programmes. The answers range from a 4-13% increase in energy use. However, during the great recession, UK energy consumption dropped by 17% and has yet to return to previous levels, let alone grow beyond them. This suggests that the affordability of energy alters our decision whether to squeeze the balloon or let out a bit of air. A European-wide study of the rebound effect made a similar finding when it observed that the amount of rebound appeared to depend on the financial situation and environmental appreciation of those taking action to improve efficiency.

We have plenty of new ideas too

Tens of thousands of scientists are seeking other ways of generating electricity or improving the efficiency of existing methods. This is just a small sample:

At the Queen Mary University, London, a research team found **noise enhances the efficiency of photovoltaics** by 40%. Although they used rock music, the serious application is positioning solar cells to absorb and convert local noise to electricity.

Staying with noisy scientists: At the University of Utah, Orest Symko leads a team of physicists exploring ways of **making sound from heat** and using it to produce electricity. While his funding is primarily military, the technology has civilian applications such as cooling microprocessors.

Many other scientists are searching for efficient ways of turning **low-grade heat** (anything below 100°C) into electricity; they are investigating a range of techniques from chemical cells (not entirely dissimilar to batteries) to new materials. These waste recovery techniques are unlikely to become major power sources, but they do help us make the most of energy. We can use them to draw out heat from steam in traditional power plants, to generate electricity in manufacturing processes and to replace air conditioning units.

You can cook a meal in 20 minutes using a **fuel-free solar-powered barbeque**.

Scientists have found that **metal particles can burn** to generate heat and be recycled (they are energy stores). Whether used to power transport or heat buildings, these particles could be the fireworks we need to celebrate our move from oil.

Off-grid communities can use a cunning device that **charges a solid-state fuel cell** in any form of fire or oven.

Wellington boots do not get as hot as an oven, but they have a large heat gradient (cold soil, sweaty feet – nice). Pop a generator in the sole and **make electricity while gardening**.

While in the garden, do not avoid the **dog mess** – scoop it up and put it in a special bin. An energy company will take it away and produce electricity.

Or **plug a light bulb into a tree** and power the fairy lights. Okay, it's a bit more complicated than that, but scientists in the Netherlands have made energy 'plants'.

What about a battery that costs around five US cents and folds to the size of a matchbox? It can power personal gadgets and, more importantly, medical equipment in remote locations. The **battery uses bacteria from wastewater**.

You can also use **bacterial spores** to generate electricity from evaporating water. You could put one in a bowl of water on a sunny windowsill and charge your phone.

If you are considering photovoltaics for your rooftop, you might want some low-cost batteries that can store enough energy to power your house for a couple of days. In April 2015, the Tesla company announced **7kWh batteries for the home**, and an industrial version capable of storing 1GWh.

Remember triboelectric charging (**rubbing balloons on jumpers**)? Scientists at Georgia Tech (an American university) have worked out a way of harvesting that energy to create a flowing current.

Why use water for geothermal energy? The oil industry squirts vast quantities of carbon dioxide into old wells to prolong production. Allowing the carbon dioxide to circulate and removing the heat could give **geothermal energy** to communities that have experienced oil extraction.

At the National Coal Mining Museum for England, engineers are developing a method to **extract heat from the water pumped** from the closed colliery that houses mining heritage.

A community scheme in Sweden **pipes the heat from a mine** to keep the homes of an Arctic village warm.

There is a lot less **heat in car tyres**, but they still get pretty hot. In March 2015 Goodyear announced a concept tyre that charges your (electric vehicle) battery using this heat.

On the surface of Earth's oceans, thousands of buoys measure the seascape. Each buoy uses energy to run equipment and to move. The boats that refuel and service the buoys also need energy. With an integral, renewable power source, these buoys would not need refuelling and would require fewer visits from their owners. Enter the Wave Glider. These crewless vessels **draw energy from waves** (and the Sun, but the wave bit is more interesting).

Kites are fun and, underwater, could draw energy from currents. In theory, they swim in a figure of eight, pulling their tether through a generator and producing electricity more efficiently than turbines. We might see some off the shores of Holyhead, Anglesey during 2017.

While moving about, why not generate electricity in your shoes, or on a walkway or even a **football pitch**.

Nuclear fusion keeps our Sun shining and releases energy by joining atoms, instead of splitting them (as in nuclear fission). It has long been the energy holy grail.[101] Serious investigations into fusion started in the Forties. The hydrogen bomb forced hydrogen atoms together, creating helium, in 1952; commercial energy production has been 'a couple of decades' away ever since. Current projects estimate we will start to buy fusion electricity between 2030 and 2050, subject to billions of pounds of funding.

The Sun manages fusion without fuss; here on Earth, it is tricky. At its simplest, fusion has three stages: ignition, chain reaction and generation. The atoms need huge temperatures and pressures to start and keep up a chain reaction, and they throw out more heat as they connect. Our old friends steam and turbine take the heat and produce electricity.

More than a hundred scientific projects worldwide are researching fusion. Some have achieved ignition, and ITER (the International Thermonuclear Experimental Reactor) is building a facility in France to prove sustained ignition (a chain reaction). No material can stand the heat of fusion, so other ways of containing the reaction are under development. The two leading contenders are lasers and magnets. The choice of this containment is the principal difference between projects. The high temperatures of fusion also mean that conventional technologies, eg turbines, need an upgrade.[102]

One project aims to fit a fusion system 'on the back of a truck'. Lockheed Martin believes it will have a commercially viable fusion cell in production by 2024.

---

101 Fusion promises to provide all the energy we need, but is not renewable as it uses varieties of hydrogen called deuterium and tritium. The deuterium we have on Earth will last only a few million years; we can, however, make tritium.

102 If you feel that fusion is futuristic, you could be right. The National Ignition Facility in the United States played the 'warp core' in the 2009 Star Trek movie.

Helium has varieties that are uncommon on Earth, but if we turn to the Moon, we can find tons of **helium-3** (it has one fewer neutron than regular helium). Should we work out a cost(energy)-efficient way of mining rocks rich in helium-3 and returning them to Earth, we could power the entire planet for thousands of years.

As we turn our attention, money and inquisitive nature to the subject of energy, we will find methods of generation we have not yet dreamt of. Our ingenuity is only constrained by time.

We need more than science and technology – communities that generate

Many ways of improving electricity supply do not rely on fossil fuels (after construction). These methods are large-scale (eg nuclear plants, geothermal and ocean thermal energy conversion); they have local or regional potential (such as wind and solar); they can produce electricity in the home or office (solar, ground source heat and micro-CHP). The variety of generation methods will change the way we think about electricity. Some people are already convinced. Let's look at some examples where communities have taken the power to generate into their own hands.

Community energy generation is popular in northern Europe and the United States, and we are starting to see more examples popping up in the UK. The number of new projects increases every week, so this section is more of a retrospective or celebration of pioneers:

Ashton Hayes is possibly the most environmentally-friendly village in the UK. The community project is the focus of academic research, government grants and private-sector support. It is evaluating the feasibility of solar- and wind-powered electricity production locally; the electricity will power the local primary school. Since January 2006, the village has cut its carbon emissions by 23%. Using our favourite assumption, that means it is using 23% less oil.

The **Brecon Beacons** in mid-Wales are famous for their outstanding natural beauty, being wet, and sheep farming. Years ago an enterprising sheep farmer, Howell Williams, installed a water turbine that extracts energy from streams on his land and uses it to produce electricity (run of the river). Because Howell can sell electricity to his local utility company and uses the grid to distribute the electricity, the enterprise works well. Lots of people have followed suit, including the National Trust. See the Sources for a photo diary of the installation of their 'hidden hydro'.

More recently, **Plymouth City Council** has set up a company to produce and supply electricity locally. The Council believes that its tie-up with a small, agile energy provider will save its residents a cool £1m a year. Using a different commercial structure, Nottingham City Council owns a not-for-profit energy company. Both local authorities see this as part of the future of electricity supply.

These examples show communities can forge changes in energy provision. But communities have another role in our story, in that large-scale power schemes rise or fall on the tide of public opinion. Communities closely involved in planning and developing are more likely to support plants during construction and are less liable to complain about them in service. However, one person's drive and vision can kick-start a revolution.

Driven by an interest in science and the wish to improve the quality of life in his village, Malawian teenager William Kamkwamba lacked the kind of infrastructure and support which made the previous (UK) examples possible. Unable to afford schooling he continued his education in the local library, and taught himself the science and engineering needed to build a wind-powered generator. In 2002 William made his first windmill, built for his family, from parts scavenged from local rubbish dumps. In the process, he gave himself and the world some (electric) shocks. Starting with local news and spread by the Internet, William's story flew around the globe. He addressed the Technology, Entertainment and Design (TED) conference in 2007, where attending entrepreneurs offered to pay for his secondary education. William is continuing his work to make life better for his fellow Malawians, and in December 2013 Time Magazine named him one of their '30 People Under 30 Who Are Changing The World'.

The numbers and diversity of examples of electricity generation and energy efficiency are astounding.

Any handful of cases must include the Polynesian island of **Tetiaroa**. After filming *Mutiny on the Bounty* here, Marlon Brando dreamt of building a luxury hotel, with local resources and powered with no fossil fuels. His dream is taking shape; the resort will use seawater air cooling, photovoltaics and vertical column wind turbines instead of fossil fuels. Also, the resort's vehicles and generators may use biodiesel (from coconut oil) and methane made from whole coconuts.

In the **northern half of our fair British Isles**, community energy schemes have enough potential to power 100,000 average homes.[103]

A design group (IwamotoScott) would like to build giant algae towers in **San Francisco** to produce hydrogen to power vehicles.

The Canary island of **El Hierro** aims to become self-sufficient in energy by using a blend of pumped hydro and wind generation and replacing 6,000 petrol and diesel vehicles with electric models.

In 2010 the government of **Puerto Rico** passed an act to diversify its electricity supply. By 2035 all electricity utilities must generate 20% of their product from renewable sources. The new law encourages a broad range of centralised and distributed generation methods: geothermal, photovoltaics, wind, biomass, hydroelectric, municipal solid waste, landfill gas, tidal, wave, ocean thermal, anaerobic digestion, and fuel cells using renewable fuels.

You might be hard-pressed to think of two communities more different than a Texan city and the environmentally rich country of **Costa Rica**, so sit down. They both make use of natural resources to power their homes, public buildings and businesses. **Georgetown** has a stated goal to use 100% renewably-sourced electricity using the abundant Texan sun. Costa Rica entered into the record books in early 2015 when it powered the whole country for 113 days without recourse to fossil fuels (again, just talking about electricity). Costa Rica produces around 97% of its

---

103 Just like the Dow Jones solar and Belfast tidal schemes we considered earlier, we have to ask whether this claim matches Department of Energy and Climate Change data, which show the average UK home using 16.28MWh of energy per year (including transport fuel). Assuming the community schemes achieve 60% efficiency for three-quarters of each day, they will produce enough electricity to offset the energy used in 87,400 UK homes.

electricity using conventional hydroelectricity and geothermal plants, with solar, wind and biomass contributing the rest.

Kenya's wind farms are among the most productive in the world (with capacity factors of 40% or more) and will sit alongside geothermal, hydro and solar to contribute to the renewable energy ambitions of sub-Saharan Africa. The International Energy Agency (IEA) believes that 45% of the region's electricity will come from renewable sources by 2040.

> *"Economic and social development in sub-Saharan Africa hinges critically on fixing the energy sector. The payoff can be huge; with each additional dollar invested in the power sector boosting the overall economy by $15."*

> *Dr Fatih Birol*

Back in the UK, the **King's Cross regeneration** installed a central CHP plant. The same plant and mechanical coolers will provide cooling. Extremely energy-efficient buildings maximise the sustainability of this development.

Yorkshire Water uses an Archimedes screw to produce power, and other water companies use the energy held by waste materials.

Water is also the basis for energy distribution in district heating schemes from **Southampton to Sheffield**. Although the companies running these systems have hot water and steam in common (a few use cold water for cooling), their systems are fuel-agnostic. Accepting various fuels lets operators adapt each system to suit local conditions. Energy sources range from natural gas to geothermal and municipal waste.

Scotland is embarking on geothermal projects; **Aberdeenshire**, better known for oil-based energy, is the site of two feasibility studies.

If you are looking for examples of storage helping communities out of fuel poverty, you need look no further than the **French Islands and Island Territories**, such as Corsica (okay, they are quite far away). These systems are still in the early planning stages (no contractor has been selected) and the French government reckons they are worth €100 million, together with the solar systems that will charge them.

You might think new technology is suitable only for modern buildings. Enter, stage left, the **National Trust**, who aim to cut their annual fuel bill by £4 million a year. That will come partly from a 20% reduction in demand and partly by installing an extensive range of renewable-energy technologies.

The **Centre for Sustainable Energy** is determined to help people in older homes benefit from energy efficiency and self-generation measures without compromising the historic value of their buildings. The first project, Warmer Bath, produced guidance for residents of that fair city. The second is broader and will guide local planning officers (as well as householders) in conjunction with English Heritage and Oxford City Council. In "Where has all the oil gone?" we will see that the Empire State Building and the White House are also benefiting from energy-efficiency measures and renewable generation.

One of the ironies of changing our energy sources is that the countries keenest to do so are those least able to afford the transition. At the UN Climate Change Conference in Paris in 2015 some 120 countries, supported by France launched the **International Solar Energy Alliance**. This platform will undertake innovative, concerted action to introduce new technologies and reduce the cost of solar.

Losing power

The case is clear: some communities are proving we can reduce dependence on fossil-fuel electricity. However, we also need to understand what will happen if we fail to follow their example. Putting it bluntly: without investment, rising demand for electricity will lead to an increase in the number and severity of the power cuts we experience.

Since the turn of the century, power cuts have affected hundreds of thousands of people. Specialists measure power cuts in *customer hours* – the result of multiplying the duration by the number of people affected. A 'large' power cut is one million customer hours (eg 1,000 people for 1,000 hours, or a million people for one hour). Considering just these big events, the worst recent year worldwide was 2011, which saw fourteen, from Brazil to New Zealand and Canada to Cyprus. Adverse weather (lightning, high winds, rain or hail) coupled with old or poorly-maintained power generation and distribution equipment caused the most disruption. Though, to be fair, some systems did not need the weather's help. If you believe the UK electricity supply is secure, mull over the following facts:

- Of the 537 power cuts reported to Eaton (a power management company) in 2014, equipment or human errors caused 287 (animals initiated three).

- In 2013 six customers in ten experienced a power cut, and the average length of those outages was 61 minutes.

- In November 2015 National Grid used a 'last resort' measure to prevent significant supply disruptions. It was the first time this measure has been used.

Electricity failures disrupt our lives in various, costly ways.[104] The cost builds up because, for example, fridges fail, letting medicines, vaccines and food spoil; water sits in pipes and treatment facilities. In hospitals emergency generators keep the essentials running, but all else stops quickly – and when emergency power runs out, critical care ends, too. Factories lose output, not only because of lost time but because when machine tools stop unexpectedly, the items they are making are wasted and the machines damaged. Fire detectors no longer work, so we have to leave buildings when power backup ends.

In hot countries and weather, air conditioning fails, leaving the vulnerable at the mercy of heat and humidity. In cold climates, heating systems fail. Tunnels close and traffic lights darken, leaving people travelling on the ground stranded in trains or caught in traffic jams. Those in the air fly on to nearby airports when local traffic control loses power. In some power cuts, looters and thieves take advantage of breakdowns in security.

Even if you have a fully-charged laptop and your data centre uses cold-water cooling, you will still suffer from the failure of local servers, firewalls and network controllers. At home, your router will stop

---

104 Figures from the Lawrence Berkeley National Laboratory show that the average customer suffers the following costs during a one-hour, summer-afternoon blackout in the US: residential $3; small-to-medium industrial or commercial $1,200; and large industrial or commercial $82,000. These values come from 2003; I leave you to work out the impact of inflation and our increasing reliance on information technology since then.

working, and without a good battery, your computer will switch off. In the UK, 90% of companies without a computer disaster plan will go out of business within 18 months of a significant loss of data. Power cuts set them on that path.

Short power cuts and brownouts have these effects. What impact will undependable oil prices have on energy investments and your life?

Chapter five

Your goose is cooked

BY THE END OF THIS CHAPTER, when you next cook dinner, pick up a takeaway or throw away a mouldy loaf of bread you will understand how these seemingly natural activities are tied up in oil. Of course, we will start with a bit of history before considering the biggest use of oil in the production of food. Then we will have dinner with people from across the globe who are loading their plates with less oil.[105]

Pre-oil food history and beyond

Hunter-gatherers had a healthy diet and way of life, so no one knows for sure why some settled in the Fertile Crescent (an area encompassing the modern countries of Iran, Iraq, Israel, Jordan, Lebanon, the Palestinian Territories, Syria and Turkey). The region had more biodiversity than Europe, which had experienced mass extinctions during repeated ice ages, and, thanks to its position between the mountains of Asia and rivers of Mesopotamia, the Fertile Crescent enjoyed greater fertility than Africa. These advantages meant early

---

105 Food does so much more that sustain us day-to-day. In 2010 30% of all working people derived their living from agriculture, and that figure is heavily weighted to the industrially developing world (for example, only 1.5% of working Brits were employed in agriculture). Yet, in financial terms, food generated little more than 3% of global added value, which we call GDP.

humans found living in one place, instead of roaming, to be successful. Agriculture took the local plants and animals and turned them into our common foods. The flora included wheat, barley, peas and lentils, and the fauna were cows, goats, sheep and pigs. Farming developed in Pakistan, China and South America at around the same time.

By planting crops instead of hunting and gathering, our Stone Age ancestors reaped more from the land. That let up to a hundred times more people live in any one area. The first farmers protected their families and communities from hunger. To complete some tasks, such as storing grain and building irrigation, our forebears co-operated with each other. Working together led to shared beliefs and practices, and surplus crops. Sharing the extra food meant some people could specialise in other activities, including commerce, manufacturing, government and law. Civilisation grew from agriculture. It's no coincidence Mesopotamia also saw the birth of writing and the wheel. As a species, we have never looked back.

As time passed, food continued to drive human development. Trading food has been one of our occupations for 4,000 years though we are more familiar with more 'recent' history. In the Middle Ages, when some people travelled to find gold and silk, others made their fortunes buying and selling spices. Most spice plants originally grew in the Far East. But trade routes passed through the Persian Gulf and the Mediterranean. Finding a path to the Far East to bypass the merchants and intermediaries of Arabia and the Mediterranean city-states motivated the European voyages of discovery.[106]

In 1488, Portuguese sailor Bartolomeu Dias was the first to sail around Africa, which he did during a storm. He could not continue to India, and on his return journey he became the first European to see the Cape (well, him, the 20 other sailors on his ship – the São Cristóvão (Saint Christopher) – and the sailors on the other two ships that also made the perilous voyage). He named the cape Cabo Tormentoso: the Cape of Storms. The name did not express the value of the route, so the King of Portugal (King John II) renamed it the Cape of Good Hope. Four years later, in 1492, Christopher

---

106 The modern country of Italy came into existence in 1861; before then, its cities formed states or republics. With Genoa, Amalfi, Pisa and some smaller cities, Venice formed the Maritime Republics, which held the gateway to the East and engaged in alliances and competition for Eastern riches.

Columbus landed in the Americas. Dias' discovery launched the empires that exploited the Indies and Columbus found a new world and new foods.

Civilisation in the Old World was born in the Middle East. However, South America had its share of religious, mathematical, cultural and agricultural developments. Thinking of food, the New World offered tomatoes, potatoes, maize and capsicum peppers (including chilli) and when Europeans started to colonise the Americas they took wheat, honey, sugarcane and farmyard animals. Historians call this swapping of edibles the Global Food Exchange. It formed part of the Columbian Exchange. Sadly, the rest of the Columbian Exchange included slaves, and diseases deadly to the South Americans, who lacked European immunity. Columbus (or his sailors) brought syphilis to Europe, and within three years of his voyage it had reached epidemic proportions.

Without the Exchange, common foods would not exist. Indian food would be mild. Ireland would never have suffered the potato famine. Europe would not need smoking bans. Every coffee brand would be African. Spaghetti would not have a tomato sauce, and Americans would not wash down breakfast with orange juice. Nor would populations have grown so large, so quickly.

We have never had so much food that we could stop developing agriculture. The need to grow more food and the wish to make money from farming led to intensive, industrialised agriculture. In Britain it started in the 16th century with enclosure, which was followed by mechanised agricultural tools. Jethro Tull's mechanical seed drill saw the light of day in 1701, and caused unrest, as farm labourers glimpsed the future of food production. Agriculture grew more industrial in synchronisation with industry. Factories provided metal and cheap tools, and mechanical equipment pushed farm workers off the land. People moved into towns and provided the labour for the new factories.

In tandem, the scientific revolution brought agricultural chemicals. John Bennet Lawes patented synthetic manure, made of sulphuric acid and phosphates in 1842. He later went on to show that nitrogen was essential to plant health. Scientists developed ways of producing nitrates for fertilisers around the turn of the 20th century. Particular players were Haber and Bosch; whose process makes fertiliser today. Haber proved his method in the laboratory in 1909 and Bosch helped scale it up to industrial volumes. They each received Nobel Prizes for their work. Until the end of World War II, most nitrates went into explosives; after that they improved farm yields.

Fertilisers tell half the chemical story of agriculture, pesticides the other. The earliest farmers deterred pests with native sulphur and poisonous plants. Arsenic and hydrogen cyanide were the first 'chemical' pesticides. Synthetic pesticides became widespread with DDT (dichlorodiphenyltrichloroethane) — again after World War II. The discoverer of DDT, Paul Müller, received a Nobel Prize for his work. Scientists and governments expected DDT to eradicate the pests that damaged crops and carried diseases. However, Rachel Carson proved that DDT accumulates in the bodies of animals and magnifies as it travels up the food chain. As it causes illness and infertility, governments banned it. Since then, other chemical families have replaced DDT and exceeded its use.

Industrialisation of agriculture lets us grow crops in increasingly hostile terrains and have increased yields in friendlier climes. Thanks to modern food production and distribution, our diets are more enjoyable than our immediate ancestors' were; for example, in the UK in the Seventies only 12% of people drank fruit juice. We have also enjoyed an enormous drop in the prices we pay for food. In 1984 the average UK household spent 16% of its income on food; by 2014, food spending had dropped to 11%.

Although we in the wealthy countries live the life of Riley, one in eight people worldwide suffer chronic hunger, and many live on the edge of poverty-related famine. Food riots in 2007-2008 made clear the cost of our inability to grow enough food to feed a growing population. As we will see in "What's the worst that could happen?", riots took place when food markets experienced several changes. Stockpiles of food reduced. Oil prices grew. Grain-producing countries experienced droughts, and the booming economies of Asia created a class of people who want to eat more meat. In combination with using food for biofuels, these causes led to high food prices and riots in Asia, Africa and South America. Food prices dropped in 2009, then started to increase immediately. Expensive food helped spark the Tunisian revolution and Arab Spring late in 2010.

Food and oil

The entire food chain depends on oil; whatever the source of your food, it has oil in its ingredients list. For each calorie of food you buy, the food industry uses five to ten calories of fossil-fuel energy getting it to you. We use more than twice as much oil in production (seed to harvest), than in transport and retail (farm to plate).

Seed to harvest

You should be able to name at least two of the top three uses of fossil fuels in agriculture. They include the machines that plough, sow and harvest, and the oil fuelling them (one).[107] Feeding plants nutrients is essential for their healthy growth, and nitrogen tops the pops – this comes from ammonia (two).[108] The next biggest use is irrigation (three). Farmers have always taken water from sources other than rain; they dug wells on every farm and diverted waterways. Modern irrigation is a bigger oil deal. All the equipment takes oil to make and pumping water is energy-intensive. Most of the rest of the oil used in food production goes to non-nitrogen fertilisers and pesticides.

Even if they do not make the top three, those last couple of items are important. To replace minerals like potassium and phosphorus in the soil depleted by farming, we mine. Potassium comes from potassium chloride aka potash. In 2014 we mined more than 35 million tonnes of potash for various uses, and fertilisers accounted for much more than half. We used about one-fifth of the 220 million tonnes of phosphate rocks mined in 2014 in fertilisers. Most pesticides we use today come from oil. Without them we would lose a third of agricultural production, and they use almost as much oil as irrigation.

---

107 For good reason, farmers do not pay duty on 'red diesel' (officially called rebated fuel) in the UK.

108 Ammonia is the main ingredient of nitrogen fertilisers. Its chemical formula is NH_3, which tells us it consists of one nitrogen atom and three hydrogen atoms. The nitrogen comes from the atmosphere, where it makes up 78% of the air we breathe. Today hydrogen comes from natural gas or coal.

Farm to plate

When we transport food, we want it to stay fresh. We already know about plastic packaging; another big chunk of oil costs comes from food processing.

We can (canning refers to cans and jars), freeze, chill, dry and irradiate food. Our efforts use oil to ensure microbes do not go wild during the journey from harvest to plate. Next time you open your store cupboard, remember the energy used to make steel cans and think on the oil content of plastic bottles and packets. Fancy bags of salad might seem a recent development, but we have used gases to protect food since the Thirties. The first was carbon dioxide, which prevented decay in fruit transported by ship. As you might expect, the technology has come on leaps and bounds, and now various gases protect fruit, vegetables, meat and seafood. Some protective gases resemble the air we breathe, with less oxygen and more carbon dioxide to slow the natural ripening of fruit and vegetables. Some food producers, eg crisp makers, pack their products in nitrous oxide (laughing gas) to make packing easier and ensure your deep-fried potatoes don't get all soggy.

You know about transport, so consider these numbers. The food we ate in the UK in 2010 travelled:

- 13,433m urban kilometres (car, LGV, HGV)
- 6,953m HGV kilometres (HGV in the UK, HGV overseas)[109]
- 25m air kilometres

From 1992 each type of transport increased: urban 26%, HGV 8% and air freight 162% (double and half again). More recently, it seems HGV and air freight are returning to lower levels – but that could just be the effect of the 2008-2009 recession. And these numbers do not include the overseas miles of our exports; we exported £19bn of food in 2013 (which is half the amount we imported).

---

109 HGV twice, this is not a double count. The first refers to HGV travel in urban areas and the second HGV travel outside towns and cities.

Fine words butter no parsnips

It's time to break the rule of not discussing politics at dinner. Let's talk about food, baby.

Meat and two veg

In almost all circumstances, meat production uses food that could feed people. When did you last catch a wild animal to eat? We use more resources to rear animals than for plant protein; the fundamental question is: how much more? An Internet search throws up many estimates, ranging from 2.5 to 30 times. Even with that variation, no one suggests meat has the same or lower oil cost than vegetables.

- The number varies with the meat you consider. For example, beef uses more resources than chicken.

- The resource you measure also changes the answer. Meat production uses water at a hundred times the rate of plant farming (a hundred is not mentioned in the range above, as you probably would have thought it an exaggeration) and meat uses 25 calories of energy to produce one calorie of food, compared with vegetable production, which uses 2.2 calories.

- The farming method affects the answer, too. Intensive, aka factory, farms use much more energy per egg than a smallholding that has enough scrubland to let its chickens roam free (ish).

We eat meat because we believe it essential to health, and we like the taste. We labour under the Victorian delusion that protein must form a large proportion of our diet. People grow fastest between birth and six months; we double in size during that period. During that time, we have a limited diet, mostly feeding on our mothers' milk, a substance containing 0.8-0.9% protein. Even as we grow older and have to heft around more weight, we only need between 41 and 49 grams for women and between 47 and 55 grams for men. The building blocks of protein are amino acids and carbohydrates. When we consume more protein than we use, our bodies store the carbohydrates as glycogen and turn the amino acids into ammonia. Ammonia is toxic, so our livers convert it into urea, and our kidneys mix it with water to create urine. Do you mock anyone for taking vitamin pills and producing the world's most expensive pee? Then you might want to check your meat intake and urea output, since, from 1961 to 2011 across the globe, daily meat consumption increased from 61g per day to 80g, and the growth did not occur in the countries which need more.

A close look at the data shows our income, rather than our beliefs, dictating our diet. The average human being uses 39% of meat and animal feed every year. This fortunate person probably lives in North Africa or the Near East. In those regions, the average annual meat consumption in 2005 was 33kg per person, with 83kg of dairy consumption. People in high-income countries (Europe, Central Asia and Australia, Canada, Israel, Japan, New Zealand, Russian Federation, South Africa and the United States) eat 80kg of meat a year. Everyone else survives on 28kg.

The chart below shows that not only do we eat more meat than our poorer neighbours, we also eat more food in total. The average daily calorie intake for the industrially developing countries is 2,619, compared with 3,360 in richer countries.

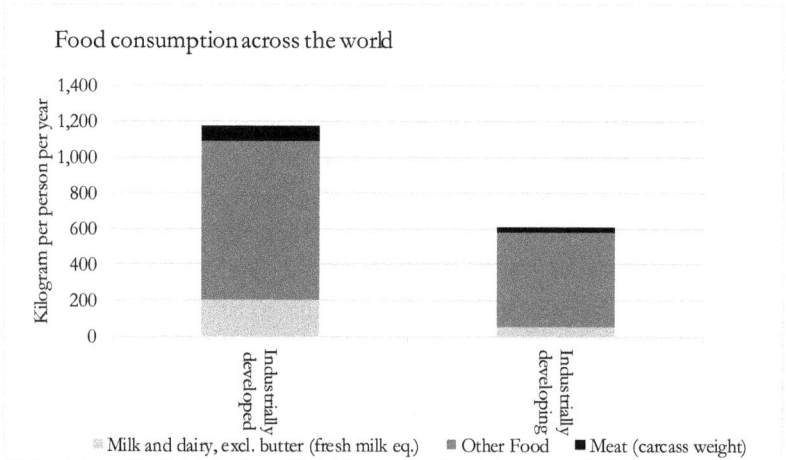

Food consumption across the world

Milk and dairy, excl. butter (fresh milk eq.)　■ Other Food　■ Meat (carcass weight)

Figure 4: Food consumption across the world

Let's recap:

† In the West, meat production uses about eight times as many calories as vegetable farming.

† We, in the richer countries, eat four times as much meat each as people in the poorer countries.

† Across the world, seven-eighths of us eat more calories than we need.

¶ We eat far more protein than we need and the fashion for meat is spreading.

We could double food production by not eating meat – possibly the least popular proposal made in the history of sustainability. Many people will more willingly use buses than significantly decrease their meat intake, so instead we see an emphasis on eating *less* meat, eg introducing meat-free days, or making the most of the resources used to produce meat, eg using every part of the carcass.

How else might we reduce the amount of meat we eat? Many science fiction authors have taken factory-grown food for granted. Isaac Asimov went one step further in his short story *Good Taste* (1976), suggesting we will become so used to synthetic foods, we will find the real stuff distasteful.

Since the Sixties, we have searched for a source of protein to feed the expanding human population. You may remember the burger made from cells grown in a lab in 2013. It cost £215,000, was 'close to meat', and proved we can grow meat from stem cells. Its makers prefer the term 'cultured meat' and many scientific teams are working on ways of making meat. We do not need to wait for its price to drop; most supermarkets sell a fungus called *Fusarium venenatum*, aka mycoprotein. Other ingredients, eg eggs and milk, turn the fungus into a textured low-fat alternative to meat. You do not need to eat these products if you're willing to give soy a chance. Tofu and tempeh contain protein, as do other legumes and nuts. However, these products can be bland; we have to go to the spice cupboard to pack some punch.

Sea vegetables (seaweed) have strong flavours – adopting them into our diets would resolve concerns about blandness. They are an essential part of diets in coastal areas, where they are available both naturally and through cultivation. So we do have alternatives to eating so much meat; we just have to change our beliefs.

Modern agriculture

You may think organic farming is about disputed claims of better taste and more nutrients. Let's shake off these misconceptions and look at the principles. Organic farming avoids synthetic chemicals by employing age-old agricultural practices, eg crop rotation or intercropping to control pests. It has also developed new techniques, such as introducing helpful insects and bacteria to keep pest numbers down. Bring on the ladybirds.[110]

Whether we farm organically or not, instead of throwing the baby out with the bathwater, we can fertilise with sewage. Using untreated or lightly-treated human waste as fertiliser spreads disease. But well-treated sewage is safe, and several countries use it, including the UK. Fertilisers made from sewage have the descriptive name 'biosolids'. Of course, when we keep animals, say on land that cannot grow arable, they provide a ready source of natural fertiliser.

Alternatively, we could stick to modern methods of fertilising the land and use renewable energy to produce the hydrogen.

In "Where has all the oil gone?", we will talk about food waste. Unfortunately, we cannot reduce some food waste – eg potato peelings – but instead of sending it to landfill, we make compost, or use it as fuel.[111]

We can reduce transport costs by producing food in urban settings. However, local food production needs to be less oily by design. Otherwise it can use more oil than farming as it lacks economies of scale. Farms in containers are self-explanatory. When you open the box, you will find new (ish) growing techniques like hydroponics and vertical farming, with the whole enterprise powered by solar panels.

Using shipping containers as space for food production opens the way for less oily urban farming, but they are not always necessary. The Gulf of Mexico oil spill in 2010 destroyed the Vietnamese-American fishing industry in New Orleans. The community responded by setting up a co-operative that trains urban farmers, sells their vegetables and fish, and secures financing for aquaponics and greenhouses. Now Vietnamese urban farmers are selling food while feeding their families. Other

---

110 To reduce synthetic fertiliser use, organic farmers plant legumes. Legumes produce pods and include beans, clover, soy and peanuts (though not vanilla pods), and host bacteria in their root nodules. The bacteria absorb nitrogen from the atmosphere and create ammonia and amino acids. The plant gets these nutrients in exchange for oxygen, carbohydrates and protein. Harvesting legumes leaves enough nitrogen compounds in the ground to fertilise later crops.

111 Individuals can go out and do a lot of the things described in *The VFUU Price of Oil*, but some need more consideration, like whether to compost or burn food waste. The most important things we can do are reduce our waste and put the stuff we can't eat (eg the tops of carrots) in our own compost or a council-provided bin.

communities, especially in America – which is light years ahead of us – also get together to grow food in urban locations, making the most of land that is otherwise unloved and untended. Other urban methods of producing food include rooftop gardens, vertical gardening and allotments.

If we move indoors, even more options become available. For instance, we can take advantage of new growing media. Soilless agriculture replaces the more conventional media with water and solids (eg plastic peanuts) or we can go one step further, we can grow plants in nutrient solution or spray it on their roots (NASA favours the latter). Growers call the methods that go without solid media 'hydroponics'. The similar-sounding 'aquaponics' adds fish and other water animals into the mix. Growing fish with food is a traditional method in the rice paddies of Southern China.

The most rock-and-roll form of food production is large urban greenhouses, stacked with shelves of food, managed by farmers with computers and without soil. These indoor farms are the supersized versions of vertical farming and hydroponics and can be found from Japan to London. Sealed buildings grow plants in artificial light. The farms use renewable energy and glass walls to reduce oil costs. In years to come, these large enterprises will complement personal and community gardens, where we will grow heavier crops and, perhaps, let a couple of chickens scratch out a living while they provide eggs.

Changing our food chain like this might seem radical until you remember sex and drugs. Usually, our characteristics pass to us from our parents through sexual reproduction. Even the tiny amoeba, which reproduces asexually, claims a clear genetic heritage. On the other hand, genetically modified (GM) plants get their strengths from other species.[112] Supporters of genetic modification say it is the next step from selective breeding and hybridisation of crops. They have shown that these foods need fewer pesticides. Opponents feel we do not understand enough about genetics to be sure we are not creating harmful foods or letting destructive genes spread through the countryside. Activists believe GM crops lead to the overuse of pesticides, somewhat negating any oil-related and health benefits.

---

112 Though dairy cows in Minnesota are now hornless, thanks to direct genetic manipulation.

And on to the drugs. Legal tobacco farming uses around 4m hectares of land (about the area of Switzerland) with a corresponding use of oil. Illegal drugs use land in countries with starving populations though we don't need to travel overseas to find resources wasted in this way. The biggest source of marijuana in the UK is home-grown. Growers use vertical farming and hydroponics to grow their profitable harvest and sell into local markets.[113] If we could replace legal and illegal drug crops with food, then we could feed more people and potentially use less oil.

Most of these solutions are oriented to wealthier countries with established agricultural industries. In poorer countries, such as many in Africa, the problems of food supply come from other sources and need other solutions. The Bill & Melinda Gates Foundation believes that African farmers will be able to feed the entire continent by 2030. They will improve yield with better seeds and by adopting new practices. They will also have access to better transport and storage so they can send food to the people who need it. Most of this will be achieved with low-carbon technologies.

Transport

We might try to do our bit and choose foods based on the journey they have made (their food miles), but this is insufficient because we must take into account each food's total oil footprint. For example, food flown into the UK may not need a heated, lighted greenhouse. Compared with UK oil-burning hothouse food, it may have emitted less carbon and used less water. If so, it used less oil.

We could reduce HGV miles by removing central warehouses, but that would add urban miles as we individually travel to farmers' markets, or as supermarkets collect from farmers in vans. A splendid answer to that difficulty is the food box. Food boxes bring fresh, often local, food to your door. Nor must it be domestic: The City Hospital in Nottingham operates a Sustainable Food Programme, in which local suppliers work together to supply farm food, improving freshness and reducing oil miles.

Fridges, while not 'transport', are essential to getting food home safely, and use a lot of energy. Supermarkets use between 10% and 50% of their energy for keeping food cold. They can reduce energy costs by

---

113 Therefore, they avoid transport costs.

putting doors on their fridges and freezers. They do not, because they think we will shop elsewhere if we have to open doors to get our sausages and chips. Bizarrely, several major supermarkets have said the government must make doors compulsory by law; then energy-saving supermarkets will not lose out to non-energy-saving supermarkets. We must garner some sympathy for our politicians as they will take a pounding when the Queen announces the Fridge Door Bill.

Food efficiency

In the UK we throw away food we could eat – 15 million tonnes of it each year. If we ate the food, we would reduce wasted oil, but at the expense of our waistlines and hearts. A better option is stop buying food we will not eat. Not so easy, eh? Or is it?

Politics

While we in the West throw away food, in the industrially developing countries people starve. While we subsidise food production in Europe and America, some countries cannot afford to develop sustainable agricultural practices. While we plan to put aside tracts of land for biofuels, bioplastics, fabrics and detergents, we cannot grow the food we need to survive. While we continue to use food resources to grow drugs (legal and illegal), we run out of the time required to set up sustainable farming in the poorest regions.

We put up barriers to decreasing the oil cost of food because we find them unpalatable. Who wants to turn vegetarian? Do you want to resist out-of-season food and forget exotic items like oranges and bananas? Does writing a weekly list of meals and sticking to your shopping list seem too much like hard work? Do you know how to take advantage of a glut of cauliflowers from your allotment or the strange-looking item in the bottom of a veg box? But if we don't adopt these practices we will find out what it means to be hungry.

Chapter six

Read oil about it

IT'S TIME TO THINK ABOUT THE WAYS we have let oil become essential to our ability to communicate. As ever, you can expect a touch of history, some facts about the amount of oil we use to keep in touch, and an overview of the problems we have to resolve to maintain our contact with the outside world.

Storytelling, giving news, and passing on experience and knowledge are important to people. Our Stone Age ancestors recounted the hunt using pictures in the caves of Lascaux – perhaps recording a map. Sharing tales of legend and mythology gave the city-states of ancient Greece common gods, history and a language. The Catholic Church encouraged travelling troupes to act out mystery plays, and hundreds of cultures have relied on an oral tradition to keep their history safe.

We started writing thousands of years ago. Since then we have recorded the epic – the stories of Gilgamesh – and the mundane – the traded price of silver and meteoritic iron. The Roman Empire issued carved bulletins and the Chinese governments of the Han and Tang dynasties published news. By the end of the 16th century, European governments had started to issue news-sheets, and within a century the first printed newspapers replaced the handwritten versions.

Today, we are less reliant on storytelling; TV, print materials and electronic communications have filled its role. Now printed media faces extinction. As forms of personal communication the handwritten letter, the greetings card and the postcard are almost defunct. Newspapers, magazines, books, CDs and DVDs are dwindling modes of mass communication. The e-revolution made possible simple and, so it seems, endless communication. All we have to do is make a single investment and pay small regular amounts. The Internet, with its wealth of innovation, took communication from professionals and gave us all the opportunity to contribute.

Is oil essential to communication? Hopefully you are already thinking along the lines of:

- Oil drives printing presses and powers the Internet.
- Plastics make CDs and DVDs, and their cases.
- Books, newspapers and magazines depend on oil in their production.
- The paper industry uses oil to grow and harvest trees and other plants.
- Oil-based processing treats paper to make it better for printing, more hardwearing and glossy.

When Gutenberg invented the printing press with movable type, he used no oil to make or power it. Now enormous printing presses have high oil costs, because of the metal they contain. They use plastic to move the paper and deliver ink while electronics control every detail of print runs. Print media uses a lot of energy – and let's not forget inks and the glues that bind books and magazines.

All forms of media use materials that travel thousands of miles. Journeys grow as raw materials move to processors and on to publishers for their final transformation. Then a distribution network kicks in, serving retailers and, eventually, us.

Newer forms of communication use more oil. We mine and purify rare chemicals to make the hundreds of parts in each gadget. Without oil, robust, attractive plastic coatings would not exist. To sell more hardware, manufacturers have embraced the ways of fashion, encouraging us to update our style with every change of wardrobe. They rely on the low cost and versatility of plastics to make new versions financially and fashionably viable. And they encourage us to replace one working model with another.

Vast amounts of energy power the computers and server farms that make up and drive the Internet, the World Wide Web and the computing industry. Considering the energy used by computers, servers, smartphones, modems and so on, we find the Internet eats 3.6-6.2% of global electricity production (84-143GW).

If we add in the energy used to make the equipment that drives the Internet and gives us access, that number jumps to 107-307GW – just short of 2% of all energy consumption. Making all the stuff to connect us to the Internet uses 53% of the energy used by the Internet (in this analysis the oil used to make plastics counts as energy).

Communications companies take a big share of the electricity bill, so they are active in their searches to improve efficiency. Individual companies like Google and Facebook make headlines when they invest in renewable generation or ingenious ways of reducing cooling costs.[114] Less well known are organisations that support energy reduction in computing. These groups look after smaller computing firms as well as companies who do not specialise in computing.

At a personal level, we waste energy so we can communicate without waiting for equipment to warm up. If you have a landline, you probably have a digital phone, which sits drawing power. A router[115] powered up day-in, day-out is equally hungry for electricity. How often do you turn off your printer, monitor, scanner and other equipment? For the cost of a multi-way socket, you can turn them all off with the flick of a switch and reduce your electricity bill by tens of pounds a year. If the value is not clear, remember: tens of pounds across 30 million UK homes adds up to the best part of a billion pounds. If we saved this energy we could turn off a fairly big power station or lots of little ones.

---

114 Google wants to use only renewable power; this proportion of their total consumption stands at 37% today. Facebook moved its servers to Sweden to take advantage of free cooling.

115 Broadband based on landlines has a built-in flaw, which the owner of the network seems disinclined to fix. Namely, if you turn off your router, your connection will slow down. This ridiculous situation does not occur with fibre broadband.

As consumers we might think we drive the Internet – after all, we do the social networking, downloading, streaming and shopping – but we are often the tail wagged by the business and government dogs.

Companies draw us into the Internet with promises of entertainment, pastimes, shopping and convenience, purely with the aim of parting us from our cash. Savvy companies use our comments to improve products and services and to influence future designs. We recommend goods and services, saving businesses' advertising budgets. Small retail businesses take advantage of the Internet's global reach to extend their market. Others open their virtual doors to new suppliers, driving down costs with 'reverse auctions'. As you might guess, in a reverse auction the price goes down as the bidding goes on. The work of service providers is easier – when done well.

Governments, on the other hand, aim to:

⟁ Reduce costs. The UK government digital strategy expects to save between £1.7 and £1.8bn a year by providing more of its services digitally.

⟁ Provide services and data. You can renew your car tax, pay your taxes and engage with the courts online. You can also find a host of documents setting out government policy – and showing how well they are doing.

⟁ Improve GDP by encouraging more sales of goods and services. In 2010 the Internet contributed 8.3% of GDP to the UK economy; this rose to 10% in 2015. In 2014 the Office for National Statistics calculated that the Internet accounted for 18% of business turnover (£492 billion). Now it is developing a measure to understand the effect of the Internet on national GDP.

⟁ Interact with voters. Every national and local government website provides information about policies and activities, and the day will come when we can vote online.

⟁ (We have to acknowledge it) spy on us.

Estonia's government has systematically invested in providing services through the Internet. These services range from voting and paying parking fees to 'attending' online university lectures and receiving medical care. Reflecting its investments, the country has a nickname: E-stonia. In South Africa the government is supplying five million digital set-top boxes to its poorest citizens. Thanks to industry pressure, these boxes will have a 'return path', meaning they will connect households to the Internet.

Internet access vs GDP per capita

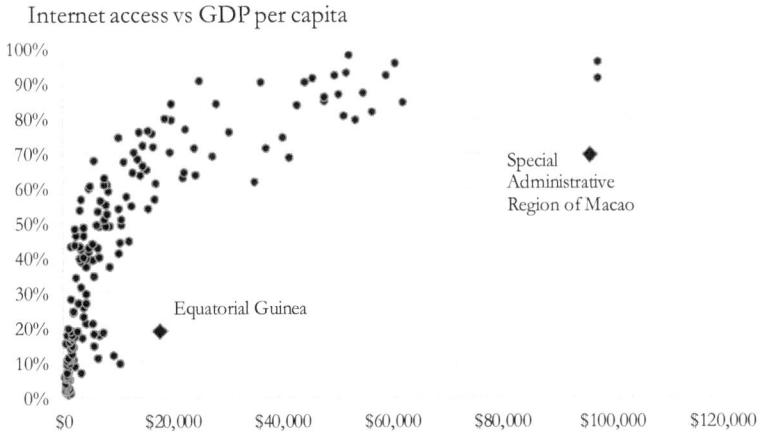

Figure 5: Internet access by nation

In the chart above, each dot represents a country. South Africa nestles in the big mass of countries with low annual GDP (<$10,000 per person) and few Internet-connected people (49%).

The chart shows that connectivity mostly relates to wealth, with a few exceptions. Equatorial Guinea has little Internet access (19%) and a relatively high GDP per person ($17,430 per person). The Special Administrative Region of Macao also has a large GDP compared to its Internet access (70% access versus $96,037 per capita GDP). Equatorial Guinea produces oil (it is the third-biggest producer in sub-Saharan Africa) and Macao hosts many gambling establishments. They are anomalies; or, if you prefer, exceptions that prove the rule. Note that some countries have a high level of Internet access, without a correspondingly high level of wealth.

59% of people around the world cannot connect to the Internet, and, for countries with low levels of connectivity, getting access is vital to improving education and economic activity. The oil cost of providing every human being with Internet access will be immense.

In the countries with high levels of Internet provision (92% in the UK), we assume the Internet is secure and will remain available. Our (false) assumption and the need to connect billions of people raises the first issue for communications in an economy with undependable oil prices: How can we ensure information flows quickly and reliably through a robust infrastructure? The vast array of choices for connectivity and the staggering rate of change bring the second: How can we reduce the oil dependency of our communication devices? The third issue is content. We know that uncontrolled communication spreads misinformation. If we use the Internet to share knowledge, we must be able to trust what we read.

Building a robust infrastructure

Even today, when fibre seems to be the norm for Internet connections, 1% of UK households (with an Internet connection) access the Internet using dial-up. Meanwhile, the government plans to provide 95% of the population with superfast (24 megabits per second (Mbps)) fixed broadband by 2017, and will give everyone the right to speeds higher than 10Mbps. Plans to add fourth-generation (4G) mobile and Wi-Fi infrastructure take the proposed reach of superfast broadband to 99% a year later. The government's role is to provide direction and funding. Business (and social enterprise) will build the infrastructure.

Will these goals put us at the top of the world's league table? In the second quarter of 2015, Romania had the fastest peak speed in Europe at 72.1Mbps. But Singapore (with 108.3Mbps), Hong Kong (94.8), South Korea (83.3), Japan (75.1) and Taiwan (74.5) beat that sparsely-populated Eastern European state. Of course, the 'average' person in those countries does not see those speeds. For example, South Korea tops the average (as opposed to peak) speed chart too, but its speeds have dropped dramatically to 23.1Mbps. On average, the world enjoys 5.1Mbps. Overall, the UK average is 11.8Mbps, which places it tenth in Europe and 19th in the world.

Do mobile communications meet our needs? 4G optimises bandwidth by moving calls to the Internet, instead of using old-fashioned telephony. Even with better efficiency, demand grows continually, and the UK government has given mobile network operators leave to use existing 2G and 3G networks for 4G traffic. When and where operators take up this option, 2G and 3G phones will stop working. Speed-wise, the average page load time in the UK is 4,724 milliseconds (ms), which sounds fast until you learn in Laos it only takes 444ms. We are the 62nd fastest in the world – so not quite in the medals.

Remote UK communities are encouraged to dig trenches for fibre optic cables, but other countries are introducing cleverer methods. For example, the 'white space' between TV channels could get the Internet to those without wires, and a trial in New Zealand has shown that balloons can provide connectivity (now if they were kites…).

The UK lags other countries, and as quickly as our infrastructure improves, other countries will march ahead. While the government and communications business have plans, they will only be good enough if we speak up for the Internet provision we need.

Making the most of our equipment

Manufacturers launch new phones, tablets and computers almost daily. Add to that smart TVs, Blu-ray and 3D players, games consoles and, as time moves on, washing machines, fridges and ovens. Functionality drives competition in communications; tablet makers want us to watch TV on a tablet while TV makers want us to browse the Web on the TV. Each gadget tries to rule the roost, and we buy items with similar functionality. For example, watching TV on a tablet is great alone, and uncomfortable in a group. Similarly, surfing on a TV is fun with friends, and eye-straining and finger-jarring when researching oil. (Been there, got the T-shirt.) We live in a constant search for gadget nirvana, replacing our technology when new functions appear – even if we do not need them.

We replace our gadgets according to fashion. Not that long ago, mobile phones shrank, at which point we mocked people with 'bricks'. Now that they're getting bigger clinging on to a small, convenient clamshell phone brings gentle ribbing from friends and family. The way we pay for phones drives an endless round of upgrades. Although the second-hand market is strong, we throw away oil every time we fall into the trap of believing new = better = necessary.

The communication drain on oil is unlikely to slow because we change our devices when they become obsolete. Switching off the 3G and 2G networks for 4G is one form of obsolescence. Another comes when businesses fail, and customers no longer receive support. However, the most common is the financial balance between the cost of supporting old products and the income they bring. As your favourite phone marches through new versions, fewer people own the original, and eventually the manufacturer stops providing updates for older phones. They stop making spare parts and stop providing guidance and support, and because you need a reliable phone which is secure against cyber-attacks, you have little choice but to upgrade.

Reducing the oil in gadgets is one of the biggest behavioural changes we can make. Unlike in other chapters, we have few good-news stories to report here. While the big manufacturers are making phones that use fewer resources and are easier to recycle, this is not enough. If we are going to use less oil in our gadgets, we need them to last longer and have support for repairs. But we are unlikely to see phones live longer while the business models of phone manufacturers rely on a 12- to 18-month replacement cycle, and we see our gadgets as disposable.

In content we trust

The Internet gives us the ability to save oil. Transport is the obvious saving; using the Internet for business and personal communications reduces the need to travel. One online-delivery round replaces dozens of car journeys. Sharing repair tips online and finding replacement parts go a long way to making our stuff last longer. But we must be able to trust the Internet.

Trust is not as difficult as you might think (steering clear of government surveillance). We need to keep software up to date, have a good virus checker and firewall, and report phishing attacks and other forms of abuse or dishonest behaviour. Most of all, we must understand that if we get something for nothing, someone has merely hidden the cost. That cost may come in the form of wrong or biased information, distracting adverts, or the personal data we give away.[116]

Communication is a mainstay of our society. Without it, we will stop moving forward and forget what we already know. Unless we solve the issues of infrastructure, disposable gadgets and content quality, we will

---

116 It is interesting that 2013-2014 saw a rash of news stories about German and Russian government departments reverting to typewriters to completely avert the risk of hacking.

not be able to continue to communicate as we do today. Do you want entertainment only from those who can afford to work for no pay? Do you feel secure in your ability to get the latest news and views? Can you get enough entertainment for long winter nights and rainy weekends?

Communications technology could be the thin brown line between another Dark Age and us. Without forethought, we might return to having to wait for the travelling minstrel to bring us news and entertainment.

Chapter seven

Homeward bound

NOW WE UNDERSTAND HOW important oil is for travelling, making stuff, creating plastics and producing electricity. And we have seen how these four uses of oil put food on your table and let you be part of the wider world without leaving your home. If only you could relax in the knowledge that volatile, fickle, unpredictable, undependable oil plays no further role in your castle. Regrettably, oil still waits on your doorstep to help you with the most private and personal aspects of your life. It is these uses of oil we round up in this chapter – together with their alternatives.

An Englishman's home is his oil

Should we describe our homes as bricks and mortar, or as oil? When we build houses, heavy equipment prepares the ground. Petrochemicals help concrete foundations set while others keep it 'plastic' (ie flexible) during its life. Plastic lumber (a contradiction in terms if you ever heard one) provides strength for walls and roofs and the frames for windows and doors. Plastic pipes bring water and gas into the home and take away waste. Oil-based paints and varnish protect non-oil materials. When the proud owners take over the house, oil carpets the floors, dresses the windows, personalises the décor, and provides the fittings in both kitchen and bathroom. If you removed the oil, the remains would not look like a home. To reduce the oil content of our homes, we must transform an entire industry.

You'll see straw bales, wood from sustainable sources, and reclaimed materials from bricks to door handles in a lot of self-build projects. The problem is, while this trail blazes low-oil materials and techniques, self-build doesn't suit everyone, and mass production relies on oil to reduce costs and keep to build schedules – which is harder (though not impossible) with sustainable materials. So, low-oil buildings often look instead to recouping the oil costs of building by reducing the oil costs of *running* our homes, offices, factories and public buildings.

'Passive' houses – those that insulate heavily and use naturally-occurring heat sources like sunlight and our bodies – are springing up everywhere. New and existing houses borrow from passive techniques to reduce the oil used for space heating, which is the biggest use of energy in most homes. In short, we can use insulation, controls, home energy production and storage to smooth demand and reduce the call on central energy production – and to recoup the energy spent building our little castles.

Here's the rub. Adding electricity generation to existing homes needs considerable investment, and building it in is not exactly small change. So, houses that produce electricity cost more and are more valuable than houses depending entirely on the National Grid. We risk letting our housing stock fall into two camps and creating a major have-and-have-not energy divide.

Moving away from building and powering our homes: From insulation in fridges and freezers to fittings in wardrobes, you'll find oil everywhere. Some companies use particle board or MDF to make furniture. Fabricated board gives two oil advantages. It is lighter than solid wood, using less oil in transport, and manufacturers make it to size, reducing waste. Its disadvantages are that it uses petrochemicals to hold together the particles or fibres and that it needs a water-resistant, oil-based coating to stop water damage.

Even the equipment in your kitchen drawer and garage are oil. When did you last see a display of wooden spoons in a homeware shop? You are much more likely to find coloured plastic, designed to accessorise your kitchen. You might have a wooden-handled chisel or screwdriver, but it is probably lonely in a box of plastic-clad tools.

We also use oil to make our homes personal. Luckily, we can substitute many decorating items with less-oily alternatives. For example, we can choose from paints and varnishes made with plant-based oils and mineral colours, metal appliances, wooden furniture and natural floor coverings. However, we are unlikely to have enough of these materials for us to continue changing home décor, furnishings and appliances with the seasons.

Dressed to kill

We don't just use oil to express our personality in the home. A quick look at the labels in your clothes will reveal polywhatnots and trade names for plastics. Yes: you dress in oil. Even if you are in 100% cotton, wool, silk or hemp, the threads holding your togs together and the dyes colouring the cloth probably come from oil.[117] Buttons and zips are plastic, as are hangers, clips and packaging. And even natural materials need oil to grow. A pair of shoes might be *to die for*, but are they to oil for?

Fashion has been around since we started to dress, marking individuality, status and wealth. In time oil replaced traditional materials, and now supplies ever-changing textures and colours. Oil enables more people to dress more fashionably, following trends season-to-season, or more individually. Fashion depends on global transport. In the time of the British Empire, producers grew cotton in Turkey and shipped it to UK factories for weaving. The fabric re-crossed the seas to Turkey for sewing. Finished clothes travelled all over the globe. Today the oil miles of clothing are even greater. Clothes sail, fly and drive – from oilfields to refineries, chemical plants, mills, factories, warehouses, shops, and, finally, your home. Transport engages low-wage economies and reduces the monetary cost of the shirt on your back, at the expense of oil. But like food miles, clothes miles are not straightforward. In this case, shopping has a big impact, and washing clothes changes the argument completely (we will come back to this).

Oil makes new, technical fibres, which allow us to do more. High-tech swimsuits helped smash world records before the authorities banned them. Divers dive deeper. Explorers travel farther into inhospitable terrains and emergency workers save lives because they have

---

117 Isn't that a misuse of the term '100%'? Yes, and it is a very common misuse.

specialist oil-based materials. And you do not need to be a specialist to be grateful for clothes that make exercise more comfortable, or offer protection in the great outdoors. After all,

"There is no such thing as bad weather – only the wrong clothes."

– Scandinavian proverb

Can we turn to natural fabrics and continue to fuel the fashion industry? Potentially not: natural materials compete with food crops, and grow in the countries that cannot afford fashion. Though cloth made from the parts of plants and animals that we cannot eat – wood fibres, wool and bamboo – have an advantage. As well as seeking alternatives to oil-based fabrics, some clothing companies are driving innovations from recyclable cotton to waste management systems – all to make the industry more sustainable.

We may choose to use oil to enable our new lifestyles with clothing that supports growing food and the more energetic alternatives to oil-based transport. We will want protective clothing and clothes that make staying healthy easy. We will need to carry clothing from specialist manufacturers to specialist users.

Cleanliness is next to oiliness

While we are thinking about using oil to look good, we need to think about beauty products and cosmetics. When you get ready for work or a night out, or brush and scrub the kids, you use oil. Personal hygiene and beauty products are oil-based, and they travel across oceans and continents to reach you. For example, shavers and sponges are plastic, and plastic packaging gets everything safely to your bathroom. Further, fancy boxes and jars express high value, saying you deserve to be pampered. We turn to mass-produced electrical appliances to get silky-smooth legs, straight hair, comforted feet and an unshaved, yet groomed, look for men. Most makeup has an oil base – including 'oil-free' foundations, which rely on our fossil friend for their ability to cover and last all day. The exceptions are the rare brands of mineral-based makeup. At the far end of the scale, we use oil to cut and pull skin, insert implants and suck out fat.

The cosmetics industry will fight to overcome the problems of undependable oil. We will turn to bars of product, more refilling, and local producers. We will remember the natural toiletries and cosmetics used by our great-grandmothers. And just as deep tans are no longer fashionable, we may find that artificial youth has had its day.

So, the dirtiest substance on Earth makes us feel young and beautiful. Strangely, it also gives us cleanliness and hygiene. Washing clothes with water is hard work; we have to scrub and pound to remove everyday dirt, and the worse stuff – eg fat – needs something tougher.

Soap has been with us for millennia, growing more sophisticated as we developed. Soap started life as a luxury item until advances in chemistry made it available to most people. As cleanliness became affordable, health improved. However, soap has drawbacks, and oil-based detergents grew in popularity after World War II. Detergents improve the effectiveness of washing and do not create an unpleasant scum in hard water. They work at various temperatures and with a variety of methods of washing, types of dirt and fabrics. Oil in cleaners gives other benefits, too. It makes fabric bright, soft, fresh and cleaner-than-clean. It produces a white, cleansing foam, and smells like summer meadows and rainbows.

Thinking about the weekly wash gives us many ideas for reducing oil use without giving up cleanliness. For example, a US manufacturer of plant-based cleaning products has calculated the potential benefits. They claim that should every US household replace one bottle of petrochemical detergent (or use one less), the unused oil (466,000 barrels) could heat or cool 26,800 homes for a year.[118] You can find many oil-free laundry products, from plant-based liquids and tablets to laundry beads[119] in shops and online. Why not give them a go? If you are not ready to go entirely oil-free, you could use concentrated detergents, which use less water and less oil in packaging and transport, as do refills and using larger bottles and boxes. You could even play with the amount of detergent you use; many people have found that they can use a lot less detergent and get the same results.

---

118 This is another situation where we have to exercise caution. Replacing oil-based detergents with plants increases the demand for palm oil. And the palm oil industry isn't the best. Disreputable growers burn down forests to clear the way for palm trees. Not only does this create smog and carbon emissions, it destroys the habitat of many animals including, most famously, orang-utans.

119 In case you wondered, ceramic laundry beads are finding a place in cleaning services and soap-free washing machines.

Another option is 'wear and air'. 75% of people believe their clothes are 'dirty' when they no longer smell the oil-based scents left behind by detergents and fabric conditioners. So, we wash our clothes much more often than we used to, using more laundry products and a lot more energy. Checking your clothes for dirt and smells, then hanging them up (if clean) can reduce your wash day blues.

One of the biggest surprises about keeping clothes clean is that 90% of the energy used comes from heating water. As more and more (oil- and plant-based) detergents are designed to work at low temperatures (30°C), we have few (good) reasons for spending oil on a hotter wash.

On our kitchen surfaces, oil gives a sparkling shine and delivers death to all bacteria. Sorry – *99% of microorganisms*. Plant-based alternatives abound; we can clean without new chemical inputs. Synthetic cloths that clean with only water work well for most surface cleaning and steam cleaning is a popular choice when items must be sterile or are very dirty. Everything that goes for laundry detergent applies to all the petrochemicals you use around the house and on your body. If you were thinking *I wouldn't want to wash with laundry beads*, look in the catalogues that take over the world in the approach to Christmas. A bestseller for many companies is a stainless steel garlic smell remover, which works in a similar way.

Like a fish out of water

To keep clean we need water. Turning on the tap not only provides clear, safe, tasteless water, but it also uses energy and other oil-based products. Water companies spend up to 5% of their running costs on energy.[120] Little wonder they have started to use the power of water and the stuff it carries to produce electricity. Water-cleaning processes also need many petrochemical materials. Modern, industrial water-purification techniques have complicated names: microfiltration, ultrafiltration, nanofiltration and reverse osmosis. These methods use plastics. The water industry also maintains and expands our sewerage and water networks. To keep it in shape they use power equipment, heavy-duty plastic pipes and many pumps.

We have not been good at managing water, and many people lack access to clean supplies. As we hunt for solutions, it is likely that

---

120 Which is up to 3% of the UK's energy usage.

desalination will become common if we can power the equipment to heat and treat seawater. You will find a solar-powered desalination method in the appendix about innovation. We can also turn to treatment processes that employ algae and generate more energy than they use.

A cure worse than the disease

People with soap and clean water enjoy better health than those without. However, cleanliness cannot prevent some illnesses. By now you won't be surprised to hear we turn to oil when we are ill. We have practised medicine throughout our history; early hominids chewed plant and animal matter to ease pain or to improve their chances of preventing or recovering from disease. Today we use sophisticated equipment and chemicals to prolong our lives.

Urbanisation changed the way we develop, make and get hold of medicines. Long ago we learned how to prepare medicines from plants and animals, and then snake-oil men made and sold patent medicines. Without science, their patients had no guarantee of a cure and ran the risk of deadly side effects. (Pennsylvanian Samuel Kier – who we'll be hearing about later – first sold oil, kerosene and petroleum butter as tonics.) With science, we learned to isolate the active ingredients in natural remedies. Coal tar provided a source of ingredients to copy naturally-occurring chemicals. Now scientists create chemical formulas on computers and build molecules from oil. They add other oily ingredients to stabilise the medicine and prevent or counter side effects. Without oil, most of us would not have ready access to medication.

Our health practices now take place in specifically built or refurbished places. They have goods made by specialist companies and are staffed by trained, educated professionals. At home, we take advantage of international development, manufacture and retail. No longer does the wise woman live at the end of the road and harvest her herbs from the fields.

Whether practised in the doctor's or dentist's surgeries, in a clinic or hospital, or at home, oil-based medicines give us historically healthy lives. We can understand the role of energy in medical practice:

- We light and heat buildings.
- We record and collect patient information and share it. This information lets any clinician treat a patient, makes our medical

services more effective, and improves our understanding of medicine.

☨ Electricity runs the machines that keep the critically ill and those undergoing surgery alive. Most hospitals have generators nearby in case of power cuts.

☨ Blue flashing lights are too familiar on our streets, and emergency vehicles are heavy and fast-moving – an oil-intensive combination for which there is not yet an alternative.

☨ Healthcare relies on routine transport to get the patient to the medical professional or vice versa.

☨ We make medicines in tremendous quantities and send them around the world.

☨ Blood and organs travel thousands of miles to save lives.

☨ During a visit to the doctor, the stethoscope chilling your skin and the needle making your heart beat faster rely on manufacturing processes and materials of high quality.

☨ Traditionally, high-quality steel made medical instruments, and sterilising baths cleaned them. Now equipment is often plastic and designed for a single use. Although that change reduces manufacturing and cleaning costs, it increases the oil we send to landfill or burn.

☨ Clever machines check blood pressure, blood sugar and cholesterol at home. They let chronically ill patients know when to ask for help.

☨ Marvellous equipment with high oil costs fills hospitals. Hopefully you have not had too much contact with anaesthetic equipment, x-ray machines, scanning equipment or intensive care wards. If you have, you will recognise that oil made them available to you.

☨ The medical industry continues to develop machines with new capabilities. For example, MRI scanners are starting to be used during operations and in the GP's surgery.

☨ Hip and knee replacements are intricate combinations of plastic and metal and give life-regenerating mobility to people with damaged joints and lost limbs. We continue to expect more and depend on oil to deliver.

⊤ Oil is the main ingredient in drugs. Paracetamol comes from phenol, phenol comes from benzene, and benzene comes from oil. (And the knee bone's connected to the leg bone.) Aspirin results from the esterification of salicylic acid with acetic anhydride. (That's enough to give you a headache.) Because penicillin and other antibiotics grow from biological ingredients, we need oil to control the temperature and mixture in huge vats and to separate and purify the drugs.

⊤ Making anaesthetics and disinfectants depends on oil directly and indirectly. For example, ethylene oxide is a sterilising gas; ethylene is an aesthetic.[121]

We have some choices for reducing our dependency on oil for health:

Reduce our use of medicine – do not take medicine unnecessarily and keep a healthy lifestyle.

Replace oil-based energy with **energy from other sources**.

Make more of oil-free transport.

Take the techniques designed for **second- and third-generation biofuels** into medicines.

Encourage innovation; for example, a team in the US have developed and shared a method of making prosthetics from PET bottles.

If we are to keep our medical standards, we need oil for anaesthesia, sterilising equipment, and treating severe diseases like cancer, heart disease and diabetes. We need operating theatres to correct disease, repair damage and help those born with disabilities. We must have hospitals, clinics, doctors' surgeries, and the ability to travel to them, or have health professionals visit us. We will cry out for medicines.

Our hospitals are clean, warm and well-lit, and we have anaesthesia. Can you image having a tooth removed, never mind your appendix, without oil?

Some examples in this chapter seem trivial while others are essential. The problem here is not the importance of the ways we use oil, but the

---

121 See the appendix "The chemistry of carbon" to learn how to recognise oil products when you see their strange-looking names.

difficulty in letting them go. Whether we are building, decorating, dressing, making up, cleaning or staying healthy, the oil habit will be hard to break.

Chapter eight

It's the economy, stupid

EVERYTHING WE HAVE TALKED ABOUT so far, you can touch. Not so with this chapter. Now we have to chew over the economy and politics. We could get fancy with both, but there are plenty of other books out there for that, so let's not. From here on in, when you read 'The Economy', think about the network of private and public sector employers, laws and money stuff that gives us employment, makes sure our spare cash grows and gives our politicians something to do to keep them out of trouble. Politics, on the other hand, is the actions taken by politicians to protect the country. And we are going to see how oil impacts both. We will also peep at the changes in the economy necessary to help us use less oil. We cover the many ways politicians can support a move away from oil in "To oil, or not to oil?".

So, the big question is, how do we use oil in The Economy? Its most significant use is as a wealth generator (like an electricity generator, but for money). Consider the North Sea; in 1964 Parliament passed the Continental Shelf Act.[122] This law made clear the legal position of the government on exploration and extraction of coal and 'petroleum' in UK coastal waters. It covered everything from licensing to pollution, and from navigation to the damage of submarine cables. You might think

---

122 Despite its name, this was not permission to build the UK's first Ikea.

that this Act opened the doors to North Sea riches, but it was only in 1969 that the first oilfields were confirmed. Before that, there were a few unsuccessful – and one disastrous – attempts to extract natural gas. Since 1975 the 100 or so British oilfields in the North Sea (which includes the Irish Sea and onshore production) have produced 25.3bn barrels of oil, and the oil industry has paid more than £316bn in production tax.[123]

Economy

In 2012 the UK offshore oil and gas industry employed 340,000 people. Of these, 32,000 worked directly for oil and gas companies or major contractors. The supply chain employed the rest, and the spending of those two groups created enough economic activity to support a further 100,000 jobs.

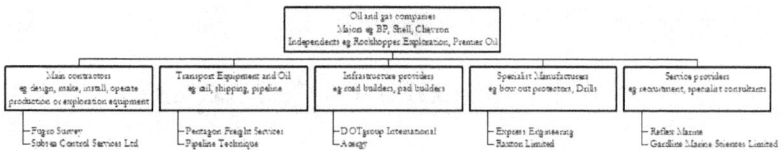

Figure 6: Structure of the UK oil and gas industry with a few examples

The 1,100 companies in the UK oil and gas supply chain sold products and services to the value of £27bn that year.

So, there's a good chance that you or someone close to you depends on the oil and gas industry for their job. And that's just the straightforward bit...

The interest on your savings comes from people who invest in business or personal debt. They only invest if they have confidence they will be repaid with interest – and that belief tends to depend on a healthy economy. The strength of the economy is measured using Gross Domestic Product, aka GDP. When more money changes hands in one month than in the previous month, the economy is said to be growing. When we keep our pennies in our pockets (or have no pennies) the economy shrinks. If you think about all the money moving around in the

---

123 Other money also flows to the Exchequer from the industry, such as fuel duty, licence fees, National Insurance and the income tax of its employees.

oil and gas industry, you can see that it helps keep GDP high and that can lead to higher interest rates.

What's more, a good proportion of your pension fund is invested in shares (or investment vehicles based on shares[124]). If your pension trustees or fund managers are adventurous, then you will have a higher proportion of shares. Cautious pension managers buy less-risky investments, like government bonds. The share prices of the UK's biggest publicly-owned companies are collected together in the FTSE 100 (Financial Times Stock Exchange Index of 100 companies). Changes in the value of their shares have a knock-on effect on the rest of the stock market, and that impacts you. Of the 101 companies in the FTSE 100 in Nov 2015, four were oil and gas companies. These four companies made up 12.3% of the market capitalisation[125] of the index. BP and Royal Dutch Shell were the third and fifth-largest companies (based on market capitalisation) in the index.

Now you know that the oil and gas industry contributes to jobs, economic growth, the interest on your savings and your pension. In other words, it is important to the economy. Before we see how we might develop our economy to be less dependent on oil, let's get our hands dirty – let's talk about politics.

Politics

We are going to see later that the oil industry is pretty old and spends a lot of money to extract oil. The resulting cost of that oil is too high for most markets. Yet oil companies pay huge amounts of tax, employ millions of people, and contribute a lot of that GDP stuff. No government can afford to lose those benefits, so virtually every government in the world provides financial support to the industry. For example, fracking was only made possible through US government investments in new technology, and to keep North Sea oil flowing when prices dropped, the UK government gave the industry new tax breaks. In many countries, governments subsidise the price of oil and oil-based products. In the UK this is done through reduced VAT.

---

124 Such as hedge funds and exchange-traded funds. And, once you claim your pension, annuities.

125 The value of each share multiplied by the total number.

To be fair, many industries benefit from government support; you saw some examples in "Oiling the wheels of industry". Other businesses benefit from lower taxes – the UK film industry, for instance. It isn't even uncommon for VAT to be reduced. So if lots of companies receive subsidies, why should oil companies be different? Perhaps because of their profits.[126]

	2013	2014	2015
British Petroleum (BP)	$23.5 bn	$3.78 bn	$6.482 bn loss
Royal Dutch Shell	$16.371 bn	$14.874 bn	$1.903 bn
Cuadrilla	£5.8 bn loss	£7.6 bn loss	

Figure 7: The profits and losses of some notable British oil and gas companies

But how much does the oil industry benefit from subsidies? Here are some numbers:

Organisation	IEA, OPEC, OECD and The World Bank	OECD	IMF	ODI
Date of subsidies	2010	2011	2013	2013-2014
Countries covered	OECD countries	OECD countries	Most countries	G20 countries
Total	US$60 bn	US$82 bn	US$4857.7 bn	US$452 bn
UK total	Not available	US$4.2 bn	US$37 bn	US$9 bn
Type of subsidy	Fossil Fuel consumption and production	Producer, consumer and general services support	Post tax subsidies	Fossil fuel production subsidies
Organisation	IEA	OECD	IMF	
Date of subsidies	2014	2014	2015	
Countries covered	Most countries	OECD countries	Most countries	
Total	US$490 bn	Not available	US$5301.7 bn	
UK total	Not available	£3.7 bn	US$41 bn	
Type of subsidy	Pre tax subsidies	Budgetary transfer and tax expenditure	Post tax subsidies	

Figure 8: A summary of fossil fuel subsidies

---

126 We need to remember that some oil companies are not as profitable as others. I will leave you to speculate how one of these companies can make such huge losses year after year.

These numbers vary widely due to a number of factors; the age of the reports is clearly one, and two further, related, factors are the value of money and the price of oil in the year of the report. And the changing constituency of the surveys doesn't help. But the most significant variation comes from the way each group defines subsidy. Some merely include the money governments pay to subsidise energy at the point of use, others include the money given to companies in the oil and gas supply chain, and then, most controversially, some include opportunity costs, like failing to decrease tax or charge the real cost of pollution and climate change.[127] Even if you pick the smallest number, you might want to consider whether this is the best use of your taxes.[128]

You could argue that subsiding the cost of fossil fuels so that we can better afford them is a good thing. However, it benefits richer people more, as they tend to use more energy and buy more goods than those

---

127 Sorry about that horrendous set of acronyms and organisations. Here are some explanations:

"The IEA (International Energy Agency) is an autonomous organisation which works to ensure reliable, affordable and clean energy for its 29 member countries and beyond. The IEA has four main areas of focus: energy security, economic development, environmental awareness and engagement worldwide."

"The mission of the Organisation for Economic Co-operation and Development (OECD) is to promote policies that will improve the economic and social well-being of people around the world."

"The Overseas Development Institute (ODI) is the UK's leading independent think tank on international development and humanitarian issues."

The G20 is a group of 19 countries and the EU, while it represents developed and developing economies it focuses on those whose size makes them influential. It seeks to make globalisation, smoother, more harmonious and sustainable.

The Organization of Petroleum Exporting Countries (OPEC) co-ordinates and unifies the oil policies of its members with the aim of stabilising the oil market. It looks after the financial interests of consumer, producers and investors.

The World Bank isn't a bank, but a collection of institutions that provide financial and technical support to economically developing countries.

"The International Monetary Fund (IMF) is an organization of 188 countries, working to foster global monetary cooperation, secure financial stability, facilitate international trade, promote high employment and sustainable economic growth, and reduce poverty around the world."

128 The ODI shows that the UK income from oil and gas was US$2.5bn in the same period.

who could fall into energy poverty. When taxes benefit the rich more than the poor, economists call them regressive. You could say the same of subsidies.

As our economy has grown dependent on oil, politicians have sought to protect supplies. We mentioned tax breaks for the oil industry a few paragraphs ago, and in "To oil, or not to oil?" we will find out how the formation of BP was linked to fuelling the Royal Navy. These actions are far from being one-offs. For instance, many people note that military action happens more often in countries with oil, and in 2012 President Obama and Prime Minister Cameron considered releasing Strategic Oil Reserves to reduce prices.[129]

War and oil

Azerbaijan saw one of the first armed conflicts motivated by oil. Towards the end of World War I, Azerbaijan and its oilfields were strategically important. So, Germany signed a treaty with Soviet Russia, giving them a quarter of Baku oil production for stopping the Ottoman Army. That battle saw the Ottoman Empire fight against Soviet Russia (the Soviet Union fought into life in 1922), with Azerbaijan forces fighting on each side. The Soviets failed. Then the Ottoman Empire fought off Britain, Armenia and the White Russians. The Battle of Baku took place during August and September 1918 and resulted in defeat for the British-led forces.

Baku was not the last battle, or war, over oil. In World War II, two critical actions had the goal of getting access to oil. In Operation Countenance, British and Soviet Troops invaded Iran during August and September 1941. The second was the Battle of Yenangyaung, which took place in Burma (now Myanmar) and saw the defeat of Japan by China, Britain and India in April 1942.

Protecting the industry also draws significant action. A very pertinent example hails from the current glut in world oil production. Exporting oil from the United States became illegal in 1975, but the opportunity to sell oil in markets with higher oil prices tempted the industry and its supporters. By the end of 2015 the ban became a bargaining chip in the US's annual budget bunfight and was lifted. At this stage, it is anyone's guess how this will impact the global market.

---

129 They didn't, but the US sold 5m barrels in a test sale two years later.

It is unlikely that reducing oil consumption will stop wars or nationalistic actions – after all, they have been around far longer than we have used oil – but it is possible for us to change our economic dependence on the black stuff, and that will change politics.

The economic alternatives

Now is the time to ask whether the alternatives to oil can keep our economy healthy. In 2010 the oil and gas industry contributed 1.1% to the UK economy. Social Enterprise (also known as non-profit organisations) contributed 2.8% in the same year. And the small and medium-sized social enterprises alone employ over a million people.

GDP growth is important. In our economy, people and businesses rely on it for income. For example, banks lend money to companies and charge interest on the money. To pay the interest and repay the capital, businesses have to increase their revenue – assuming all other costs remain equal. Similarly, people who trade in stock[130] wager on growth. The bet pays off if share prices increase or shareholders receive dividends. Growth gives investors the confidence that share prices will grow, and without growth it is hard for companies to consistently make profits, without which they cannot pay a dividend.

Of course, social enterprises don't pay dividends. They put money back into their business and the social or environmental situations they want to improve. We must not confuse social enterprise with socialism. Such organisations compete on the open market and the state takes no role in their management. Social enterprises take many business structures. The most familiar are mutuals (eg Building Societies), Provident Societies (eg BUPA) and co-operatives (eg the Co-operative Group). More recently a new type of registered company – a community interest company – has grown as a form of social enterprise. And there is a broad range of examples, such as Co-Wheels, The National Lobster Hatchery, Plymouth University and The Shakespeare School Festival. Social enterprises need not be small. The John Lewis Partnership

---

130 What is the difference between stocks and shares? Both refer to the ownership of a part of a company, but stock is a generic term and shares are held in a specific company. So you might say you own some stock and mean specific shares in one or more companies. (Or you might like the wonderfully scented flowers...)

employs 88,700 full-time owner/ employees and the Nationwide Building Society has 850,000 employees.[131]

Our society regards growth as the primary, if not the only, measure of economic success. So ridiculing change is easy, though foolish. You may know the saying 'you get what you measure'. We can understand this well by thinking about factories that pay people for each part made. At the end of the day, you might expect many parts, but not necessarily the right quality. Those factories also have to define, measure and reward good quality. In the economy, focusing on growth leads to growth[132] and leaves the quality of life to chance.

> *"Too much and for too long, we seem to have surrendered personal excellence and community values in the mere accumulation of material things. Our Gross National Product... counts air pollution and cigarette advertising, and ambulances to clear our highways of carnage."*
>
> *— Senator Robert Kennedy, March 1968*

For Kennedy, GDP[133] measured "everything in short, except that which makes life worthwhile". If Kennedy leaned too far to the left for you, it is worth bearing in mind that Simon Kuznets, who developed GDP, also felt it was only a partial measure of the success of an economy.

Growth in GDP does not reflect improved living standards. For example, after the 2008-2009 recession, the financial measure told us the economy recovered late in 2013. At the same time, many people were in lower-paid, inconsistent employment. By valuing other measures, such as median income and the cost of living, we can drive actions that benefit more people and our natural world. We can also address the financial calculations that skew our actions to the wrong goal. For example, GDP increases if patients spend more on healthcare than they lose through reduced income. Economists, politicians and authors suggest various alternative measures:

---

131 There are also interesting alternatives in the form of community currencies and time banks.

132 Roughly half the time.

133 Gross Domestic Product and Gross National Product are similar enough for us to use them interchangeably in this book. If you want to get more technical, a full description is only an Internet search away.

Median income

Cost of living

Health

Longevity

Child mortality

Greenhouse gas emissions

Natural capital (eg acreage of natural landscape, number of species)

Circular economy index[134]

Many national governments and the United Nations are introducing such measures. Including our own.

Changing the basis and measures of the economy will take time and be difficult, but, in the words of Scottish philosopher Thomas Carlyle:

> *"If something be not done, something will do itself one day, and in a fashion that will please nobody."*

---

134 What's that now? Champions of the circular economy advocate ways of supplying and consuming goods that do not start with raw materials and end with waste (as in the 'linear' economy). They are driving changes that will reduce waste, through reuse, renewable energy, composting, sustainable materials and services. They believe these changes make financial, as well as environmental, sense; for example, one study states that the UK economy could grow by £29bn a year (1.8% of GDP) by not importing as much raw material.

Part Two

Do we need to use less oil?

IN THIS PART OF THE BOOK we will think about the ways that the volatile, fickle, unpredictable, undependable price of oil impacts our lives.

But oil is a fascinating subject, full of derring-do, so we start with a little bit of geology and history in "Where did all the oil come from?". Because this book is not about bashing the oil industry, we pose the question "What has the oil industry ever done for us?", and explore the work and creativity that goes into every barrel of oil. Bearing in mind that we all use it, we will also look at the risks the oil industry takes on our behalf.

We ask "What influences the price of oil?" and take a peek at the vexed subject of "Peak oil" from a balanced perspective. In Part One we looked at the ways we use oil, and you will have already worked out some of the impact undependable oil prices will have on your life. In this part we will pull it all together and ask "What's the worst that could happen?".

Finally, we look at what happens to oil after we use it, in "Where has all the oil gone?".

All of these chapters simplify the subject of oil so that we can take the conversation out of the offices of the elite and under our control. Don't be afraid to put your toe in the water – it's lovely.

Chapter nine

Where did the all the oil come from?

IF YOU EVER WONDERED HOW ALL THAT OIL got underground, or how we found it, this is the chapter for you. Even if you haven't thought about it before, you will find out how the oil industry started and get a feel for the entrepreneurial skills that came into play when we humans chose to use oil.

The fossil fuels take far longer to form than the lifetime of any creature. Yet life provides the raw materials. Plankton is a catch-all term which describes tiny sea creatures. These specks of life include bacteria, tiny plants and the larvae of sea creatures. Plankton occupy the bottom of the food chain. Every animal living in the sea eats them or eats something that eats them. Plankton that avoid being dinner die and settle on the ocean floor. There, they mix with mud. As time passes, the layer of plankton and mud builds up. If the layers build up quickly enough, hardly any oxygen gets caught up in the plankton and the organic material cannot decompose into carbon dioxide.[135] The growing layers build up pressure, which with movement in the materials – for example, through subsidence – increases the temperature. The plankton turns into a rich

---

135 Lots of factors affect the amount of oxygen buried with the plankton – for example, sea level rises and ocean currents. Sometimes bacteria help to use up the oxygen, but if there are too many bacteria they reduce our plankton to carbon dioxide.

soup of organic material called kerogen. The mud forms source rocks. These rocks hold the kerogen and provide the high pressures and temperatures needed to create hydrocarbons, which are simply chemicals made of hydrogen and carbon atoms. Tens of millions of years pass in a few sentences.[136]

The temperature of rock beds varies with depth and pressure and fixes the blend of hydrocarbons. Natural gas forms between 150°C and 200°C. Longer chains assemble at lower temperatures (50°C to 150°C) and make oil. Below 50°C, kerogen remains unchanged. Few people dispute that oil is made this way; however, some sources do not depend on (terrestrial) life – these are called abiogenic.[137]

The rocks that form around the hydrocarbons influence how we extract them:

- As oil and natural gas are buoyant, they rise through permeable rocks. If they hit an impermeable layer of rock or salt, called a seal, [138] it holds them in place. The oil and gas stay in the permeable rock, which we call a reservoir. Reservoirs are a little like a sponge (bath, not cake), as the cells within the rock hold the oil. They hold conventional oil and gas, and we get them out with conventional drilling.

- As oil and gas move into reservoirs, they create upward pressure. The pressure may create gushers when we drill into the reservoir and lets us start extraction with no more effort than drilling. However, if the source rock has minuscule pores it would take

---

136 The oil we call conventional typically formed in the Mesozoic and Palaeozoic Eras, about 60 to 540 million years ago, and the oldest bitumen deposits (in China) date from around 1.4 billion years ago (the Mesoproterozoic Era).

137 Two principal theories seek to explain the origin of abiogenic oil. The first says cosmological methane was trapped in the Earth during its formation and seeps ever upwards, changing into more complex hydrocarbons as it goes. The second states the comets and asteroids that pummelled the planet during its early years brought carbon and carbon compounds, which moving tectonic plates force into hydrocarbons. Proponents of the abiogenic theory believe it exists in abundant quantities. If they are right, we need only worry about the costs of extracting and using such reserves.

138 Or 'cap', in the US industry.

more pressure[139] to get the oil or gas out and, stubbornly, they stay in place. 'Tight oil' is the name for oil trapped in source rock. We call gas stuck in a similar predicament 'shale gas'. We use fracking to free them.

⸙ Oil and gas that do not hit a seal escape to the surface where either they evaporate or bacteria consume them. Most oil and gas has gone this way.

⸙ Younger reserves lie close to the surface where bacteria can reach them and consume the lighter oil. They leave the heavier oils (eg bitumen). Heavy oils do not flow or rise and we have to use heat (to make them flow) or mining to extract them. We extract kerogen using the same methods.

On the other hand, most coal formed in the Carboniferous Period, about 299 to 359 million years ago. The formation of coal and the creation of oil and natural gas have three main differences.

⸙ Coal is not buoyant; we have to bring it up to the surface.

⸙ Complex plants make coal. Three to seven feet of plant matter make one foot of bituminous coal. With more heat and pressure, anthracite forms and has more energy per tonne.

⸙ All three fossil fuels form continuously, but coal needs a particular set of conditions to get started. The plants have to be dry enough to grow, yet they must be covered in water when they die to exclude oxygen and prevent decay of carbon compounds. Swamps, rising seas and subsiding land provide these conditions.[140]

The most significant differences between types of oil are their density, viscosity and sulphur content. Until very recently we depended almost entirely on one type of oil, so we call it 'conventional oil'. Conventional oil has a low density and floats on water. It has a low viscosity and flows easily. Although medium and heavy oils float on water, they have higher densities than conventional oil and flow less easily. Some count as

---

139 You know this from playing with sponges in the bath. The sponges with big holes (pores) quickly release their water when you lift them up, whereas sponges with small holes have to be squeezed and even then they hold onto water.

140 When you read "Where has all the oil gone?", you might realise that we are recreating the conditions to make coal by burning so much of it.

conventional, because we extract them in the same way. On the other hand, very heavy oils – for example kerogen and bitumen – do not float on water, and, most often, they do not flow like liquids.

So much for density and viscosity; what about sulphur? Oil is rarely only hydrogen and carbon because other substances connect to the hydrocarbon chain. The most harmful interloper is sulphur. With more than half of 1% sulphur, oil becomes sour. Oil with less sulphur is sweet. Each oilfield produces a different type of oil, and they tend to have straightforward names like Brent Crude (originally from the Brent oil fields of the North Sea, it is light and sweet) and West Texas Intermediate (American oil, which is also light and sweet). However, some have unexpected monikers like Mars Blend (American, medium and sour) or Troll Blend (Norwegian, medium and sweet).

Light, sweet oil is the easiest and cheapest to extract. It is easier and cheaper to refine. Its cheapness made it our first choice until production levels peaked.[141] Many remaining oil reserves are very heavy or have high levels of sulphur. In the main, only offshore (including the Artic) and tight, tricky locations harbour new reserves of light, sweet oil. In this book, the term 'conventional oil' refers to light-to-heavy oil extracted from conventional wells, including offshore.

The history of oil and humans is interesting; let's take a quick peep at it before talking about the ingenuity that goes into every barrel of oil. Evidence of the first use of hydrocarbons comes from 40,000-year-old Neanderthal tools found in Syria, which have bitumen stuck to them. Not surprisingly, China lays claim to some fossil-fuel firsts. They piped natural gas in bamboo pipes to fuel a salt factory around 500BC. In the same facility, more bamboo pipes took brine to the salt works, where burning gas heated the liquid, evaporated the water and produced salt. In this early industrial use of methane, Chinese engineers had to prevent explosions. They controlled safety by mixing the fuel with air in the world's first carburettor. The Chinese also dug the first oil well, in 347BC.

Farther west and several centuries later (the 13th century, in fact), Marco Polo found a thriving oil industry in the modern town of Baku,

---

141 According to Dr Fatih Birol, at the time Chief Economist of the International Energy Agency (IEA) and now its Executive Director, conventional oil production peaked in 2006. Or, to put it another way, conventional oil extraction was at its highest in that year, and has dropped ever since.

Azerbaijan, where the streets were paved with bitumen. Historians believe Polo described Baku in these terms:

> *"Near the Georgian border there is a spring from which gushes a stream of oil, in such abundance that a hundred ships may load there at once. This oil is not good to eat; but it is good for burning and as a salve for men and camels affected with itch or scab. Men come from a long distance to fetch this oil, and in all the neighbourhood no other oil is burnt but this."*

Other travellers reported the oil industry before Polo. The Window on Baku website claims, at least, four previous travellers visited the area.

Other early interactions between humans and fossil fuels

Bronze Age Britons used coal in funeral pyres, and the Romans exploited many British coal fields to heat their baths. Our ancestors got more than energy from fossil fuels. In Persia, India and Greece, religious sects thought flames fed by natural gas were divine. The Native Americans and early Europeans waterproofed their boats with bitumen, and the Ancient Egyptians used it to make mummies. The Aztecs decorated with coal (as did the Romans, Chinese and Victorian Britons).

The difference between ancient and modern uses of oil lies in the amount and variety of ways we use it today. Now the oil industry employs millions of people, oil travels farther than any substance on Earth, and we depend on oil for our comfort and welfare.

Several[142] men could claim, if they were alive, to be the father of the modern oil industry:

> David McKain, an oil industry expert and historian, asserts that as early as the 1820s **West Virginians** drilled for oil and used it for lighting and lubrication. He believes Edwin Drake (who we will meet shortly) used West Virginian oil to keep his drills turning.

---

142 As with other inventions and discoveries, several people simultaneously asked the same question and found similar answers. For oil, the key question was 'Can oil replace and improve on common fuels?'. Maybe you can name the competing inventors of television, the telephone and the incandescent light bulb.

In Azerbaijan (in the same Baku region visited by Marco Polo) the Russian state controlled the oil fields and exported oil to Persia for lighting. In 1846 the engineer **Alekseyev** dropped a sharp blade to create the first mechanically-dug well in the region. The **Nobel brothers (Ludvig and Robert,** not Alfred of dynamite and Prizes fame) set up an oil company there in 1879. At one point Branobel (Nobel Brothers) was one of the world's biggest oil companies and the largest in Russia.

Abraham Gesner[143] was a Canadian doctor with an interest in geology. He turned his hand to developing a fuel to replace whale oil, as the latter burnt with a lot of smoke and was expensive. His success came in 1846 when he heated coal and distilled the resulting vapours to produce a clear liquid. He called the liquid kerosene, [144] which means wax-oil. He installed kerosene lights in towns in the United States.

Gesner was the first modern scientist to make kerosene from fossil fuels. However, he failed to apply for a patent, and the Scottish industrialist **James Young** beat him to the punch. Young distilled kerosene from oil shale. He used a rock known locally as boghead coal, and scientifically as torbanite. He successfully applied for an American patent in 1852. Young also refined oil seeping from a coal mine in Riddings, Derbyshire in 1848.

In 1853, **Ignacy Lukasiewicz**, a Polish pharmacist, found a way of extracting kerosene from oil. He made the first kerosene lamp, which lit night-time surgery at his local hospital. Lukasiewicz also funded an oil well in the Carpathian region of Poland in 1854. He continued his affair with oil by opening Europe's first oil refinery

---

143 Gesner (1797-1864) was a Canadian of German descent, who agreed to study medicine to win the hand of the woman he loved. While studying surgery at St Guy's hospital in London, Gesner befriended Charles Lyell, a well-known influence (and friend) of Charles Darwin. He returned to Canada with a keen interest in geology. Gesner surveyed Western Canada, where he found coal and fossils.

144 Kerosene is a mix of hydrocarbons; we change its composition to suit our needs. Not all countries use the name 'kerosene', though – for instance, UK and Asian markets know it as paraffin. Chemists use 'paraffin' to mean a waxy blend of hydrocarbons. While Gesner was the first modern scientist to extract kerosene, the Persians used a similar method to extract 'white naphtha' from coal and bitumen.

in Ulaszowice in 1856. He was a modest and generous man and gave away his expertise (even to Rockefeller's Standard Oil) as he felt he was working for the benefit of humanity – not for money.

Some writers credit **Oil Springs, Ontario** as the first commercial oil well in the world, with a drilling date of 1858. However, the Oil Springs website makes it clear this was an accidental discovery.

In Pennsylvania, **Samuel Kier** found black liquid polluting his father's salt wells. Kier first tried to sell distillates of the liquid as tonics or petroleum butter, [145] without success. Then he turned to science and his efforts paid off: he safely extracted kerosene from the oil. His method produced cheaper kerosene than other sources. When George Bissell, a New York lawyer, heard about Kier, he asked Yale Professor of Chemistry Dr Silliman to find out whether they could make money from oil as a source of kerosene. Silliman's positive report helped Bissell find financial backers for the 'Pennsylvania Rock Oil Company'. The lawyer hired 'Colonel' Edwin Drake to drill an oil well in Titusville, Pennsylvania in 1859. Kier went on to set up refineries and distilled oil from Drake's well.

We should heed the words of Gesner:

> *"The progress of discovery in this case, as in others, has been slow and gradual. It has been carried on by the labours, not of one mind, but of many, so as to render it difficult to discover to whom the greatest credit is due."*

Although we cannot definitively assign credit, we can see that, over a short period and in several places, the three factors needed to kick-start the modern oil industry came together:

---

145 Petroleum butter is not an unappetising version of peanut butter, but something like petroleum jelly. After Kier had given up on his early attempts to sell oil and turned his attention to kerosene, Robert Chesebrough visited the Titusville oilfields, where he noticed workers applied 'rod wax' to their cuts, scrapes and burns. The rod wax healed the wounds. Rod wax is a waxy residue that formed around the pumps at the drill head. The workers removed the wax and noticed its 'healing powers'. Chesebrough extracted petroleum jelly, which he called Vaseline, from the rod wax and the rest, as they say, is history. Now we understand that petroleum jelly moisturises the skin, which promotes self-healing. It also forms a protective barrier, allowing natural healing to take place. The Chesebrough Manufacturing Company made Vaseline until 1987, when Unilever bought the company.

T Realising oil could provide a cheap, liquid fuel.

T Learning how to extract oil.

T Developing safe, high-volume methods to distil or refine oil.

Oil in the UK

The UK oil industry has a long and distinguished history. For example, the East Midlands sits on a bed of oil-bearing rocks, and in the 19th-century oil from local collieries flowed in the streets of Alfreton (the same oil distilled by James Young). The UK sank its first mainland oil well in Derbyshire in 1919. When the Government took ownership of mineral rights in 1934, the Duke of Devonshire kept the ownership of oil under his lands; he is the only person in the UK with these rights. The 'The American Roughnecks of Sherwood Forest' is a little-told story of World War II (the link is in the Sources). It is well worth a read.

From entrepreneurial starting points, the oil industry grew, integrated with the economy and became more creative. Kerosene soon replaced biofuels for lighting, heating and lubrication. Then engineers put another distillate, petrol, into the internal combustion engine. Soon after fuelling transport, we developed plastics. Plastic billiard balls replaced ivory and plastic spectacle frames replaced tortoiseshell. The more we use oil, the more ways we dream up for using it. The demand for oil wavers during times of economic hardship, but our craving has never ceased. And the oil industry works hard to satisfy our needs.

Chapter ten

What has the oil industry ever done for us?

I BELIEVE MANY PEOPLE SHY AWAY from the subject of oil because it tends to provoke intemperate language, conspiracy theories and the all-too-prevalent habit of tarring everyone in the oil industry with the same brush.[146] At the end of this chapter you will find the recognition that the oil industry has inbuilt hazards, which are not always managed successfully; however, for the most part, we celebrate the hard work and ingenuity that gives us all the things described in Part One. Here you will follow oil's journey from unknown reserve to useful product. You will also learn about different types of oil, including tight oil and shale gas. Have fun.

Thanks to slick transport, refining and management, few of us have seen oil in its natural state, or ever will. Pictures of nodding donkeys and perilous oil platforms cannot bring to life the true scale and difficulty of the oil industry, nor its influence on your life. Oil is dirty; it smells; it pollutes. It is full of stuff we don't want, like hydrogen sulphide, nitrogen

---

146 Or, in the words of a friend, "It's all a bit heavy, man". (Okay, he didn't say "man", but I bet if he'd known I was going to quote him, he would have.)

oxides and other hydrocarbons such as benzene. People and oil do not mix well. It burns, poisons and destroys what we love about our world.

Given the dangers of oil, we should ask two questions: Why do we use oil? How do we win this unique substance from the bowels of the Earth? Part One showed 'why'. The 'how' also deserves attention. So, let's take our hats off to the men and women who make all the advantages of oil available to us and minimise its effects on our world.

The first commercially-extracted oil was easy to find, extract and move. Anyone could start an oil company, leading to the oil rushes of the late 19th century. As the business of exploration and extraction grew harder, the companies consolidated. 'Consolidated' is the polite way of saying the larger or richer companies bought the smaller, less fit, or unlucky ones. With their deep pockets, the oil companies fund developments in chemistry, geology, transport and engineering and continue to advance the industry. Now we extract oil in hostile environments, carry it far and wide and turn it into a growing range of useful products.

The oil industry employs hundreds of different professions and reflects the world at large. It also has its own skills, language and culture. Geologists find reserves, and engineers work out how to extract every drop of oil from a reservoir. Armies of accountants work out how to make the most money from it. Chemical engineers turn 'black gold' into useful chemicals like petrol and the raw ingredients of plastic. Logisticians manage billions of barrels of oil at sea, on the road, travelling by rail and flowing through pipelines.

The rest of us support the experts: From advertising 'mad men', to supermarket checkout staff; from the factory workers making consumer goods, to the retail staff who sell them. Between us all, we keep the global economy ticking.

Despite where we find oil or what type it is, the basic journey travelled by each drop is the same:
- Exploration finds oil.
- Extraction removes oil from the ground.
- Transport takes oil to the refinery.
- Refining separates hydrocarbons, removes impurities and transforms the oil into useful stuff.

☥ Other industries use oil and, eventually, we get our turn (which we covered in Part One).

There is a sixth stage - decommissioning

Once an oil well stops producing oil, it has to be taken out of service and the area around it restored. The vast majority of UK oil wells are at sea and they are slowing in production. This slowing is sufficient to have created a new sector within the oil industry which focuses on removing and disposing of platforms and securing well heads. Decommissioning in the North Sea cost £800 million in 2014. The price tag is expected to fall just short of £17 billion by the end of 2024 and £40 billion by 2040.

Exploration

The oil industry has recorded every oil find since the early days of commercial oil extraction. It knows every possible detail about that oil and its geological prisons. From these data, specialists predict the location of further oil reserves. New reserves provide more data, and in turn we unearth more oil. With infinite conventional oil, we would have a virtuous circle. However, as we venture into less-accessible reserves, we need experts with supercomputers to find the places where oil lurks.

As we saw in "Where did all the oil come from?", oil flows up through layers of rock and geology sometimes traps it with seals of impermeable rock. Geophysicists use the latest technology to understand possible oil-bearing features. They work out where the features sit, the rock type, and their size. Then we have to check whether their calculations are right.

Seismic exploration is a common way of exploring oilfields.[147] It is similar to the echolocation used by bats, but instead of noise, the geophysicists create shockwaves with pounding equipment on land and air guns at sea. When the waves move between layers of rock they bounce back and cause new vibrations. Scientists measure these 'echoing' tremors with geophones (like microphones for shockwaves in rocks). At sea, they lay geophones on the seabed or tow hydrophones behind ships. (That's right: hydrophones are like microphones for shockwaves in water.) Seismic exploration creates vast amounts of data. Software analyses the data and turns it into images, and specialists examine the

---

147 Seismic exploration is good, as it is much cheaper and less intrusive than drilling.

images to identify 'plays'. 'Play' is the industry term for an area containing source rocks, a reservoir and a seal.

Before they will pay for extraction, investors need physical confirmation of the money they can make. The oil industry uses a measure known as Technically Recoverable Resources to report how much oil they can extract. Going one step further, Economically Recoverable Resources (ERR) describes the reserves that could make profits for oil companies and investors. The amount of ERR depends on the cost of extraction and oil prices. To show that a reserve will be profitable, the would-be producer drills an exploratory well.

Exploratory drilling uses the same technology as extraction, and the costs soon mount. Onshore or off, this is an expensive and risky business. You can expect to spend upwards of $10 million per well, with no guarantee of success. Dry wells are sealed without producing (enough) oil. The company loses the money spent on exploration and may even fail to recover its licence fee.

If a well is productive, then scientific, engineering and financial wizards analyse the data from the well with the goal of getting the money to start extraction. They may recommend using exploratory wells for extraction to reduce costs.

Extraction

The first stage in oil extraction is creating a well and putting equipment in place. The second stage is oil production.

Digging the well and building the production rig

Early people dug through rock with hand-driven chisels, until the Chinese invented a machine to lift and drop a sharp blade. With this technique, called percussion drilling, they created salt and brine wells thousands of feet deep. Edwin Drake used percussion to break up the rocks of Pennsylvania though steam engines did the work of lifting the blade. Percussion drilling was interrupted by regular breaks to let workers take out broken rocks, so Drake invented the drilling technique still employed to create oil wells today: he used a rotary drill to make his way through the rocks that sealed the oil reserves. A pipe followed the drill, protecting the well from collapse and giving the oil a clear path to the surface.

What has the oil industry ever done for us?

How low do they go?

Early oil wells were no more than a couple of hundred feet deep. Advances in drilling technology have pushed wells more than ten miles (16km) away from the point the drill enters the ground, and to great depths. The deepest – the Odoptu Op-11 well (based off the Russian island Sakhalin, North of Japan) – reaches down more than seven miles (12km). The depths of Odoptu Op-11 are hotter than 200°C. Not surprisingly, the same field boasts the world's largest production platform, which can extract 12,000 tons of oil a day (that's 3.2 million barrels a year).

But drilling is a bit more complicated than putting a hole in the ground. Drilling mud (a mixture of oil, rocks and chemicals) lubricates and protects the drill. It also stops the hole collapsing. Concrete (called casing) packs the gap between the pipe and the surrounding rocks. Gravel or holes at the bottom of the casing let the oil flow into the well.

Perhaps the biggest drilling innovation is horizontal drilling. Drilling away from the vertical avoids areas not suitable for drilling, like soft rocks. It increases the oil extracted, as reservoirs often lie horizontally, and allows better placement of the wellhead. Horizontal drilling makes fracking in shale profitable.

Our final major drilling advance is drilling into the seabed. Various writers make claims for the first offshore well: Was it in 1846, at the Bibi-Eibat field in Baku, or in 1896, when the Californian Summerland Field expanded into the sea, or in 1892, in Grand Lake St Mary's (Ohio)?[148] Offshore drilling opens oil reserves on the continental shelves and opens an area roughly the size of North America for exploration and exploitation.

Working offshore has two big advantages: low human population and more unexploited reserves. At the same time, it poses more difficulties than on land, and that adds to costs and risks. The team that drill the well install a ton of kit around the same time. To extract as much oil as possible safely requires sensors, valves and masses of computing power. At some point, all wells need encouragement, and pumps draw the oil

---

148 Why mention inaccurate claims? Well, it is amusing when various websites claim firsts. A little research rapidly uncovers a more varied, interesting history than otherwise suggested. Bear this in mind if you decide to poke around the fascinating world of oil for yourself.

through the well or inject gas or water to stimulate production. Blowout controllers stop the pressure of the oil rupturing the well. Pipes take care of local transport. Equipment at offshore wells removes the water and salt mixed with the oil.

Flareconomy

Most oil comes with natural gas, which disrupts the flow of the oil – a little like the carbon dioxide in a shaken pop bottle.

Operators need to dispose of natural gas. If they can transport gas to market, it delivers a handy income. However, sometimes oil exploration takes place in remote locations with no means of moving the gas. In these cases, the cost of selling the gas can outstrip income – so operators flare, or burn, the gas at the well site.[149] All wells have equipment to separate oil and gas; most also have furnaces and chimneys to manage flaring.

Satellite photos make clear the extent of the North Dakota and Texas oilfields. The photographs highlight land disruption and gas flares. In the Bakken oilfields (North Dakota) at the start of 2014, operators did not sell 40% of the natural gas they produced. The oil companies burned the bulk of it, and some escaped.

Some countries have either banned flaring or impose fines to reduce carbon dioxide and other emissions, and because it wastes a valuable resource.

➢ *Yet in 2013, the world flamed away 4.9 trillion cubic feet of natural gas.*

➢ *That is 4% of global gas production and about one and a half times as much as we consumed in the UK.*

➢ *In the UK in 2014, our oil industry, including terminals, flared or vented 37bn cubic feet (a working week's consumption). Flaring produces all the pollutants linked with natural gas, without harnessing its energy.*

---

149 The numbers given here include 'venting'; that is, letting the gas escape to the environment. Remember this when you read about the Global Warming Potential (GWP) of methane.

> *The industry does make some attempts to tackle flaring, by changing its working practices and selling more gas, but the amount flared will likely increase as production increases.*

All this equipment takes up a lot of space and needs constant oversight and maintenance. It needs a stable electricity supply, which adds substations, cables and control systems to our list of gear. Remote sites may have electricity generators. On land, the rig spreads across the well site, and multiple wells in the same area share equipment, such as pipes to collect the oil and take it to the transport system. Onshore wells present several choices for locating equipment and only suffer the usual vagaries of weather. Offshore extraction brings a host of extra challenges.

At sea, the distance between the rig and the wellhead presents the first issue. Even at shallow depths, where platforms attach to the seabed, water pushes on the drill and pipes and moves them. In deep water, floating platforms or ships hold the rig, and waves, tides and currents batter the floating equipment, moving it in relation to the well. These movements further increase the need for control equipment and space to house it.

To overcome space constraints and the dangers of extreme weather and icebergs, the industry is putting increasing amounts of equipment on the seabed. Professional divers set up and look after underwater equipment at shallow sites, but they cannot go into deep waters. Remotely Operated Vehicles (ROV, or Rovers) support human divers at all depths. And from 300 feet below, they boldly go where no person can and perform the tasks people cannot.

Moving equipment to site is also trickier at sea

No one can tow one of the massive oil platforms you have seen on TV. Instead, the oil industry uses enormous, semi-submersible lift ships to move awkward stuff. These craft sink by filling their ballast with water and they manoeuvre their decks until they're below the equipment, ship or platform they will carry. Then they expel the ballast water to return to sailing buoyancy, picking up the equipment, ship or platform as they rise.

Once the equipment is in place, tested and proven extraction starts. Producing the first barrel of oil is a momentous milestone. To get to that point has taken six to ten years, and has cost billions of pounds.

Production, aka recovery

Each reservoir carries several wells, to maximise extraction. At the start of production, the natural pressure of the oil pushes it out of the wells. This phase – primary recovery – releases between 5% and 15% of the oil in a reservoir. Over time, the pressure falls, and the oil stops flowing. Then producers convert some wells to secondary recovery to urge the oil to flow.

In general, the secondary recovery phase uses one of two methods: pumping the oil, or increasing the pressure of the well by injecting in air, natural gas, water or other fluids.[150] When the secondary phase stops producing oil, we have recovered an additional 20-30% of the oil. To maximise the oil taken, we move on to tertiary extraction.

The characteristic feature of tertiary methods is that they reduce the viscosity of the oil remaining in a reservoir (making it runnier).

- In some reservoirs, heat loosens the oil.
- In others, detergents break up the oil (like cleaning grease from pots).
- Texas producers use microbes to split molecules in the oil; the resulting lighter hydrocarbons flow more easily.
- A standard method is injecting water followed by carbon dioxide. The carbon dioxide reduces the oil's viscosity and the water pushes the oil into the well.
- Many wells are subject to hydraulic fracturing.

Tertiary recovery only releases up to 20-30% of the reservoir. Altogether, traditional production techniques extract little more than half the oil in a reservoir, and costs increase with each phase of production. Operators only use the latter stages if the selling price is high enough. Sometimes work at a reservoir with Technically Recoverable Resources grows too expensive. Then the owner mothballs the wells until rising prices improve the financial outlook.

The gas industry also returns to old wells. For example, Statoil has recently built and sunk a platform the size of a football pitch in the Norwegian offshore gas fields. The enormous and expensive piece of kit will extract gas from a reservoir financially 'depleted' a few years ago. The

---

150 Injecting carbon dioxide into old oil wells could create a CO_2 store. This method is undergoing tests in the North Sea.

particular geology of a play can also make old wells more financially viable, as some depleted wells 'recover' as oil rises from source rocks and refills the reservoir.

Transport

The two common forms of long-distance transport are pipelines and tankers. Pipelines form a complicated network; wells and fields supply the pipeline, which in turn feeds a distribution network. Pumping and control stations push hydrocarbons along, manage flow and look for leaks. No one can enter the pipelines and they often lie underground or underwater, so autonomous units called 'pigs' inspect, clean and repair the pipes. Alternatively, inspectors fly over pipelines in helicopters or use drones to hunt out leaks, and sophisticated instruments detect corrosion in the pipe walls. Pipelines cost less than road or rail transport for fixed, high-volume hydrocarbon journeys.

Two financial factors slow the use of pipelines. First, the rapid depletion of smaller fields could stop a pipeline seeing enough action to justify investment. Second, operators delay pipeline construction until they are confident that a new find will pay for itself. In both cases, rail is a common alternative.

Politics and the environmental impact of pipelines also drive transport by rail. The most obvious example is the infamous Keystone XL pipeline. After years of planning and debating, the US government has declined permission for the last link in the pipeline that would take Canadian bitumen to the refineries of the Gulf. Supporters of the pipeline promise to keep fighting; in the meantime, the oil will continue to travel by train until it meets with the pipeline in Nebraska.

Oil tankers are a relatively cheap and reliable way of transporting oil by sea. The birth of the tanker predates the modern oil industry; in the mid-19th-century dockworkers poured oil from Upper Burma into the hold of ships. Oil tankers have come on since then. The world's first recognisable oil tanker entered service in 1878 with the Branobel Company. The *Zoroaster* was filled from a specially-built pipeline and boasted safety features which were doubted in the 19th century but are now commonplace. Each year, oil tankers transport about 60% of the oil we use. This is almost 21bn barrels.[151]

During 2002 and 2003, Daewoo built the world's four largest oil tankers. Two are now Floating Production, Storage and Offloading (FPSO) vessels[152] (oil storage tanks) working in the Middle East oilfields. The remaining ships (*TI Europe* and *TI Oceania*) each carry more than 3m barrels – half a billion litres or 200 Olympic-sized swimming pools – of oil. Both could host three football pitches laid end-to-end. They are fast, managing a blistering 16 knots fully laden. That's about 19mph.

Oil-based chemicals help improve tanker safety and speed. Coatings inside the tanks reduce 'hydrocarbon release', and, in turn, inert gas protects the coatings from oxygen. Oil tankers are often white to reflect the Sun's rays and to prevent temperature increases in their volatile cargo. Shortly we might see nanotechnology-coated oil tankers as scientists develop new methods to reduce water resistance.

Logisticians control transport and make sure the right amount of oil reaches the refinery at the right time. Too much oil drives up storage costs and ties up cash. Too little and the refinery runs short. Shutting down a refinery is a big deal because the owners will face a period with no income and significant restart costs.

Refining

Refining separates hydrocarbons and converts less-valuable materials into more-profitable ones. Oil contains many types of hydrocarbons, and no two oils are the same. Not only does oil vary from field to field, one reservoir will produce several mixtures. However, the users of oil products need only a few hydrocarbons and they must be sure that they'll get exactly what they want. Understanding how to release useful hydrocarbons safely has been a black art since ancient cultures first used

---

151 A barrel of oil is exactly 42 US gallons: just shy of 35 imperial gallons and a touch less than 159 litres. The oil transported by sea each year would fill 1.3m Olympic-sized swimming pools.

152 FPSO vessels are common and as prone to size inflation as everything else in this industry. For example, Shell is investing in a massive vessel which will produce 5.3 million tonnes of liquids per year (most of which is LNG). It was launched into Australian waters in November 2013.

oil. Only those inducted through long study and practical experience understand the art and science of refining – the author claims none at all.

Refining has many stages, starting with the mundane acts of storage and cleaning (to remove water and salt) before the excitement begins. Refining proper begins with distillation. When we distil alcohol from fermented vegetable juice, we get a single usable distillate. With oil, the refinery yields parts or fractions, so we call it fractional distillation.

Distillation sees the oil heated in stages until it reaches a temperature between 330 and 370°C and moves to an atmospheric distillation column. We call it 'atmospheric' because the column pressure matches that of our atmosphere at ground level.[153] The top of the tube is hotter than the bottom. The high temperatures encourage the lighter hydrocarbons to evaporate, so they rise and separate. Pipes, set at intervals, collect the fumes, each a different hydrocarbon (or distillate). They travel to condensing chambers where they return to a liquid state. The evaporation leaves extra-thick stuff at the bottom that does not budge at the temperature and pressure of the atmospheric distillation column.

Higher temperatures are more expensive and less safe, so the gooey mass moves to a vacuum distillation column. This second column has a low pressure, so the otherwise inert gunk evaporates. Again, temperature increases with height, encouraging the hydrocarbons to rise, separate and leave the column.

The refiner removes all sulphur before the next stage of refining. (Sulphur is not nice. It is often present in the form of hydrogen sulphide.[154] Left in fuel, it makes acid when burnt. It also damages the catalysts needed to break down heavy hydrocarbons.) Sulphur bonds with the oil, and another hot job, hydrogen desulphurisation, prises them apart. At temperatures between 300 and 400°C, a catalyst encourages the sulphur to change places with a hydrogen atom. (Catalysts are chemical agents that encourage other chemicals to react. They help reactions without undergoing any permanent changes themselves.) We do not

---

153 Not because it is a good place for parties – it isn't.

154 This gas smells of rotten eggs, is highly flammable and will kill instantly at concentrations as low as 1,000ppm (0.1% by volume).

waste the sulphur.[155] The same chemistry makes the trans fats found in some processed foods – yum.

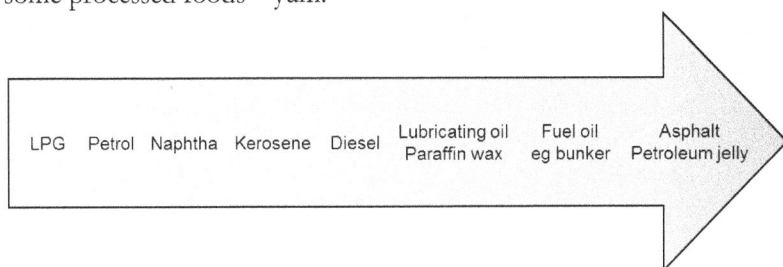

| LPG | Petrol | Naphtha | Kerosene | Diesel | Lubricating oil Paraffin wax | Fuel oil eg bunker | Asphalt Petroleum jelly |

Figure 9: The distillates of oil – lightest to heaviest

If we line up the hydrocarbons produced so far according to their weight, the most valuable sit at the light end. To sell every drop of oil at the highest possible price, refiners break the long chains into something lighter and more valuable: most often, transport fuels.

'Cracking' is the group of methods that split heavy hydrocarbons.[156] The cracking type used depends on the fuel you want. For example, India uses more diesel than petrol and employs 'hydrocracking'. In this method, hydrogen and noble metal catalysts (eg gold) break the carbon bonds in long hydrocarbon chains. Similar to hydrogen desulphurisation, it makes clean diesel and kerosene.

If you need petrol, then you will turn to 'fluid catalytic cracking'. You need a mineral called a zeolite, which you heat until it becomes fluid – now you know how the method got its name – before adding heated heavy hydrocarbons. (Zeolites are a family of minerals called aluminosilicates; they are made of aluminium, silica and oxygen (and other bits and bobs). They are porous and help divide carbon chains, like pushing tomatoes through a sieve.) After the cracking, steam removes the hydrocarbons from the zeolite, and burning removes carbon deposits

---

155 Some uses of sulphur: as fertiliser, to treat spots, in detergents, to vulcanise rubber, to dry fruit and to bleach paper.

156 In the diagram we can see how the distillates of oil line up. We have met them all except naphtha. Naphtha has two primary uses: it is either used as feedstocks to make petrochemicals, or is cracked and reformed to produce high-octane petrol.

and heats the mineral ready for reuse. The hydrocarbons go on for further distillation.[157]

Petrochemical engineering

Distillation and cracking are two forms of petrochemical engineering. Petrochemical engineers design, develop and manage the equipment used to make valuable compounds from oil. They create chemicals from fuels to plastics and satisfy the needs of millions of industrial customers. They maintain the safety and efficiency of the refinery; environmental impact; and quality.

We have reached the end of conventional oil's journey. Formerly-inaccessible, unpleasant oil is now valuable chemicals ready for use in manufacturing, medicine, farming and transport. Unconventional oils take a similar route to your door.

Develops unconventional oils

We saw in "Where did all the oil come from?" that the production of conventional oils has already peaked (that is, reached its maximum and started to decline). Other methods of producing oil now meet our growing demands. The oil companies make liquid fuels from coal or natural gas. They extract and refine unconventional oils, such as heavy oils and bitumen. They heat oil shale to get at heavy hydrocarbons and crack rocks to release light oil. Let's take those methods one at a time.

In the 1920s, coal-rich and oil-poor Germany encouraged scientists working at the Kaiser Wilhelm Institute to **create liquid fuels from coal**. The first step of their solution was making carbon monoxide from coal with the method called gasification.[158] For the second step, they created the Fischer-Tropsch process. Fischer-Tropsch is a series of chain reactions that combine carbon

---

157 All those hot jobs may have you thinking that the oil industry must have big fuel bills. It does. The UK's oil and gas industry uses about 4% of our energy consumption, of which refineries use about half.

158 Gasification turns the hydrocarbons in fossil fuels or biomass (eg cellulose) into carbon monoxide, carbon dioxide and hydrogen. In the UK we used it to create town gas before we started to exploit the gas fields of the North Sea. Most gasification processes heat the hydrocarbons and force them to react with oxygen or steam. Gasification produces syngas, which we burn to generate heat and electricity, or turn into other fuels.

monoxide with hydrogen to form hydrocarbons. Companies from Qatar to South Africa to the United States use Fischer-Tropsch, and other gas-to-liquid techniques, to create liquid fuels out of natural gas and coal.

Because unconventional oil does not flow easily, operators use steam, electricity or fire to loosen the **kerogen** in oil shale deposits and force it to flow to the surface. Freezing the ground around the shale keeps the liquid kerogen moving up, rather than sideways.

Tar is a thick liquid made from coal and the heavy hydrocarbon in tar-sands looks like tar, but it is **bitumen**. Operators use steam or solvents to loosen the bitumen, before pumping it out of deposits.

Heavy oil producers also use traditional surface-mining techniques to remove bitumen- or kerogen-rich rocks and sand. They crush or heat the minerals to release the hydrocarbons. **Bitumen and kerogen need a lot more processing** than conventional oil to make the valuable, clean, light hydrocarbons we use, so changes to refining go hand-in-hand with non-conventional methods.

Oil and gas get stuck in source rocks, where low permeability stops them flowing easily. The industry developed **'hydraulic fracturing'** to get light oil from these previously unworkable reserves. We know it by other names, like fracking, oil fracking and tight oil extraction.

Confusing oil shale and shale oil

Let's have a quick definition check. 'Oil shale' is kerogen captured in shale; we extract the kerogen by heating or crushing the stone. 'Shale oil' is the light oil held in shale and other impermeable rocks. Shale oil is also called 'tight oil'. (Because 'tight oil' is a touch more accurate and is less easy to confuse with oil shale, that is what we call shale oil in The VFUU Price of Oil.) Only fracking frees tight oil. As well as opening some reserves of tight oil, fracking increases the oil extracted from conventional wells and revives old wells.

Fracking

The history of tight oil and shale gas extraction

Returning to the 19th century, we see early fracturing methods in use even before Colonel Drake dropped his first drill. The modern gas industry started in Fredonia, New York in the 1820s. Fredonia had the first gasometer, metered supply and natural-gas street lighting. Remember, the UK used town gas (made from coal). A gas well in Fredonia experienced the first use of cracking (though Appalachian gas producers happened on the technique around the same time). The method was much more spectacular than modern hydraulic fracturing. Preston Barmore dropped gunpowder into a gas well, followed by a hot poker. Happily for him, the force of the explosion did not damage the well, him or others above ground. It did, however, create cracks in the rocks and a greater flow of gas.

Barmore's cracking took place between 1857 when he drilled the well in question, and 1859, when drilling starting in Titusville. In the 1860s oil well owners in the hydrocarbon-rich states of Pennsylvania, West Virginia, New York and Kentucky used solid nitro-glycerine to break oil-bearing rocks. In 1865 a Civil War veteran, Colonel Roberts, applied for a patent for an oil well torpedo. His successful application gave him a monopoly on explosive fracturing. The technique was superseded in 1949 when a Kansas oil well experienced hydraulic fracturing near to the form used today.

The oil industry uses fracking everywhere; indeed around 60% of all oil wells are fracked. By the Seventies the UK, Netherlands, Germany, other parts of Europe and North Africa had used fracking to stimulate wells. In the same decade, the United States hit its peak of conventional oil production. The government, keen to keep the oil industry alive and protect America's energy supply, funded research to improve fracking and increase domestic oil and gas production. The technical skills, along with government and private funding to make tight extraction profitable, were pulled together by George Mitchell, whom the oil industry hails as the father of fracking. And in 1997 Union Pacific Resources developed 'slickwater fracturing'. This technique used the advantages of horizontal drilling to open tight reserves of oil and gas. Fracking became cheap when compared with conventional oil and gas prices.

The name *hydraulic fracturing* describes the process exactly. High-pressure water fractures oil- bearing rocks to give our bounty a path back to the well. The term 'slickwater' comes from the fracking fluid, a cocktail of water and chemicals which help the fracturing and extraction. First of the additives are proppants: sand or other hard particles that keep open the cracks. Proppants settle out of pure water, so guar gum and borate salts increase and maintain the viscosity of water, and push the proppants along. (Viscosity is like gooeyness, but less sticky.) Other chemicals, from acids to polymers, increase production. Some improve the well condition and can provide other benefits; for example, in some wells radioactive compounds show the progress of cracks.

Fracking changes the entire industry, as it massively increases the amount of oil and gas we might be able to extract.[159] For example, proven conventional- and tight-oil reserves amount to 40bn barrels in the United States. There are a further 78bn barrels of unproven (but technically recoverable) tight-oil reserves (of course, there are also unproven reserves from other sources). Tight oil could make the United States into the world's biggest producer of oil for several years. Globally, tight-oil reserves could deliver 418bn barrels. We also have 7,576 trillion cubic feet of unproven, technically-recoverable natural gas reserves worldwide, enough to fuel the world for 62 years. At this point you should be wondering if the UK can benefit from shale gas exploration – see the appendix "Is the UK's shale gas all it is cracked up to be?" for more details.

Many countries look to shale gas and tight oil for their financial and political benefits. Successful production would boost their energy supplies and would offer independence from other oil-producing countries.[160] No wonder the oil industry wants to get fracking.[161]

---

159 The world's oil (and gas and coal) reserves are split into two camps. 'Proven' reserves are those we are fairly certain we can extract. On the other hand, technology, leasing arrangements, regulations and financial uncertainty mean there are other reserves which are 'unproven'.

160 There are also benefits for the environment, as these light oils and natural gas produce less carbon dioxide and other pollutants for each unit of energy produced.

161 Unfortunately, the benefits of shale gas are likely to be short lived, for example the production of the four largest US shale gas fields peaked between 2011 and 2015.

The economic benefits of fracking

The oil industry has seen substantial jobs growth in the United States, with benefits trickling into local communities, even for those without directly relevant industry skills: the hotel trade has boomed, and the largest national growth in work for women has been waiting on tables.

Lower industrial gas prices have given the United States economy a boost after the recession of 2008. A global gas market could replace local markets, which would give gas-rich countries – for example, the United States – export opportunities. International trade will reduce gas prices in some countries and raise them in others.

Fossil-fuel riches in the United States have the potential to cause seismic shifts in political power. Attendees of the 2013 World Economic Forum in Davos talked of a swing in economic power from the Middle East to America. Majid Jafar, a Middle-Eastern oil expert, made three statements in an interview that few would have predicted a few years earlier. First, he said that the MENA (The Middle East and North Africa) region must become efficient in its use of oil (like the United States in the Seventies?). Second, he called for energy policies based on strategy, instead of politics. Third he warned that it could take as long as 50 years for natural gas to supplant oil.

Shortly after Majid Jafar made these statements, Prince Alwaleed bin Talal warned that Saudi Arabia must diversify to overcome its economic reliance on oil. He referred to the thriving fracking industry in the United States, and elsewhere, when he made this statement. You will find examples of MENA countries turning to renewables in the Sources.

Attendant risks

The oil industry works in the most challenging environments on Earth. Extraction equipment faces enormous depths, high pressures and extremes of temperatures. Only computers can test this equipment before use in the field, and computer simulations are never exact. Every installation has fail-safes and backups, but it is impossible to second-guess the exact conditions they will meet. The industry learns the risks of new extraction methods to our collective cost. Starting with the infamous BP explosion and spill in 2010, here is a small selection of the incidents that have occurred in the current decade:

April 2010: An explosion on the Deepwater Horizon semi-submersible offshore oil drilling rig kills 11 men and dumps more than 200m gallons of oil into the Gulf of Mexico. The clean-up operation recovers no more than 20% of the spill, and five years later over 10m gallons of oil are found on the seabed. Tighter regulations to prevent a recurrence have yet to make the statute book in the US.

July 2011: An Exxon-Mobile pipeline ruptures under the Yellowstone River in Montana. Between 750 and 1,000 barrels of oil are released in 30 minutes. Less than four years later, a pipeline owned by Bridger Pipeline leaks 30,000 gallons into the same river.

March 2012: Corrosion in a well casing on the Total Elgin platform releases between two and 23 tonnes of condensate (which lies in a six-nautical-mile-long sheen on the North Sea) and up to seven million cubic feet of methane a day. The platform is flaring gas when the leak occurs.

March 2013: A Shell platform (Kulluk) in the Arctic loses its mooring and runs aground on Kudiak Island. This incident is similar to an incident in July 2012 when locals reported the Noble Discoverer (a Shell drilling rig) had run ashore in the Aleutian Islands.

July 2013: The Koh Samet beach in Thailand is drenched in oil when the pipeline feeding oil to a tanker bursts. State-owned oil company PTT Global Chemical plc owns the pipeline.

During 2013: The Canadian tar fields experience four 'releases' of bitumen. They spill for several months. Eventually, Albertan authorities order operators CNRL to drain a small lake to find the source of the leak. Several months later, conflicting statements say either 1,878 or 1,177 cubic metres (11,812 or 7,403 barrels) have been spilled.

July 2013: A runaway train carrying heavy oil explodes in downtown Lac Megantic (a Canadian city), killing 47 people and flattening a large area of the city.

March 2014: In Galveston Bay, Texas, an oil barge collides with another vessel and spills 168,000 gallons of bunker.

April 2015: In English Bay, Canada, the M/V Marathassa discharges an unknown quantity of fuel oil. The clean-up took 16 days.

May 2015: A broken pipeline spills over 100,000 gallons of oil. 21,000 gallons reach the Pacific Ocean off Santa Barbara County, California. It kills wildlife from seals to birds and blots a previously beautiful stretch of coastline.

May 2015: In Texas an oil well explodes and releases poisonous quantities of hydrogen sulphide and other chemicals. 20 families have to leave their homes; some may never be able to return.

November 2015: Two trains derail in Wisconsin. The first spills thousands of gallons of ethanol into the Mississippi and the second releases oil, which is contained before it reaches a waterway.

Winter 2015-2016: An underground natural gas storage facility starts to leak in October, attempts to plug the leak succeed in February. In the meantime, local residents report all manner of health impacts and many families relocate.

February 2016: A broken pipeline in Peru spills 3,000 barrels of oil, contaminating the rivers supplying water to local communities.

Throughout the decade: A set of wells in the Gulf of Mexico start to leak in 2004 and are still pushing out oil. Unplugged these wells will leak for another 100 years.

The Presidential report into the 2010 Gulf of Mexico fire and spill stated, in its closing chapter:

> *"... The root causes are systemic and, absent significant reform in both industry practices and government policies, might well recur."*

The opportunities and hazards of keeping our oily habits affect us all. The oil industry and oil-producing countries continue to invest billions in developing new methods of finding oil and extracting it from increasingly harsh or less-profitable locations. They develop new technologies, such as cold cracking or extraction using microwaves, to reduce their own energy bills and the cost of oil products. They also work hard to recover more oil from spills.

Our governments support the status quo through subsidies, legislation and political actions. For example, the UK government seems willing to risk political strife and military action to continue exploring and developing oilfields around the Falklands, and the US government has given permission for exploratory offshore drilling in Arctic waters off Alaska.[162] These actions make possible (and increase the cost of) oil's journey from the reservoir. The same governments seek to minimise environmental impacts, such as oil spills and methane leaks; the regulations they impose on the industry will increase costs (which is why the industry lobbies against them). Perhaps creating a vicious circle?[163]

This chapter recognises the work the oil industry puts into fuelling our economy. Remember, we pay them to undertake these actions. If we drive a petrol or diesel car, use disposable plastic products or buy imported out-of-season vegetables, we would be guilty of a touch of hypocrisy in criticising the oil industry and failing to recognise the demands we place upon it.

---

162 While Shell have backed out of the Arctic, there is nothing to stop other operators taking up their licences.

163 You could even say viscous circle.

Chapter eleven

What influences the price of oil?

THIS BOOK DOESN'T PREDICT OIL PRICES. Search the web or social media and you will find plenty of people willing to give that a try. But we do need to put a toe in the murky waters of oil prices, and understand how a host of factors make oil prices – and therefore oil – undependable.

A spider could have woven these topics together; pull one thread and the others move. Rather than get caught up in endless, circular arguments, we will look simply and briefly at each subject.

Oil price influences – a spider's web

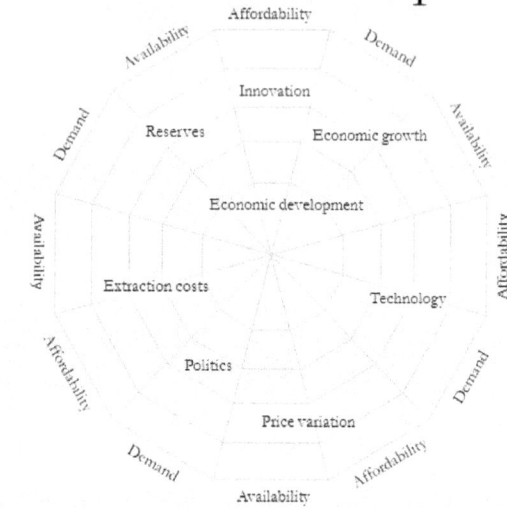

Figure 10: The factors influencing oil prices

Availability – reserves

Reserves are the oil we have yet to extract. We already know they have been in the making since the Mesoproterozoic age (1.4bn years ago). We know that they fill up with the carbon from tiny creatures and abiogenic sources. We know that all that life, death and chemical whizzery-pokery has put a lot of oil beneath our feet (especially when our feet are in boats). However, you are unlikely to know *how much* oil lies in these reserves. There are many organisations which research the amount of oil in the Earth and publish data about 'proven' reserves. Proven-reserve data tell us, with 90% confidence, how much oil we can economically recover. The data used for the estimates come from several professions, from geologists to economists. This table shows the estimates made by three respected organisations in 2015. The differences in the amounts are explained by the different ways each body defines oil; for instance, BP includes natural gas liquids.

There are a couple of cautions that come with these data. First the amount of oil left under each oil-producing country varies massively. The chart below shows the reserves by country; according to this dataset the UK and another 80 countries each have less than 10bn barrels of oil:

Data source	Energy Information Agency (EIA)	Organisation of the Petroleum Exporting Countries (OPEC)	British Petroleum (BP)
Amount of oil (billion barrels)	1656	1492.88	1700.1
2014 annual production (billion barrels)	33.980405	26.7983365	32.365645
Amount of oil (years)	49 years	56 years	53 years

Figure 11: Proven oil reserves and how long they will last

The second thing to remember is the amount of 'unproven' reserves. These data estimate the oil we know about, but cannot be sure we will recover. As technology improves and oil prices grow, these reserves become easier and more profitable to extract. As a result, they may move to the 'proven' group.

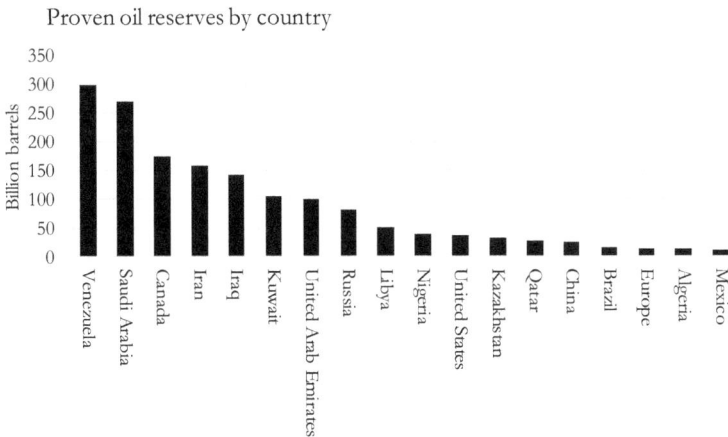

Figure 12: Proven oil reserves by country

2014 was a weak year for oil discoveries; what's more, this was the fourth year in a row that discoveries fell. A research firm called IHS stated early in 2015 that only 16bn barrels of oil were discovered in 2014. One way of thinking about this is that all the money spent in 2014 only turned up half a year's worth of oil. The US Energy Information Agency (EIA) were slightly more optimistic. It says proven reserves increased by

7bn barrels during 2014. However, OPEC only raised its 2014 estimate by 0.2% in 2015.[164]

Proven reserves by region and year

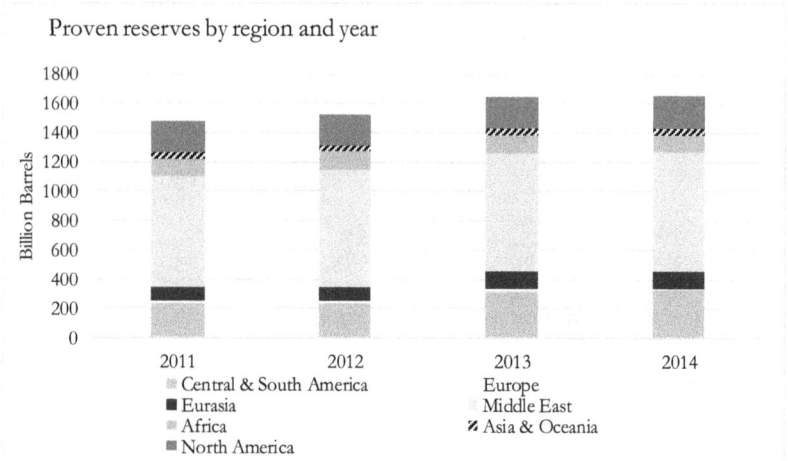

Figure 13: Proven oil reserves by region and year

Availability – technology and innovation

Once upon a time, extracting (and refining) oil was child's play. It lay close to the surface, on or near to land, and in porous reservoirs. Now we are seeking out oil deep below ground, far offshore, in tight reservoirs and from oil sands. We need new technology to find and extract oil.

New technology comes with three major costs: the development of the technology, the payments made to investors and the steps taken to minimise risks. Investors only stump up cash if they are convinced the market will bear all these costs. So, when the price of oil drops, fewer people are willing to invest, and both technology development and exploration slow. The flip side is that new technology can cause an oversupply situation. When US shale oil producers massively increased supply the resulting glut was the primary cause of price falls in 2014-2016.

---

164 The IHS figure shows new discoveries – when the oil industry says 'wow, we didn't know that was there'. Some of these discoveries may be proven, but most of them were not. The EIA and OPEC figures show how much unproven oil became proven during 2014.

Availability – price variation[165]

In addition to driving investment, price has another impact on availability. If the price is low, some producers will stop operations. Stopping production can be expensive to do, and halts all income, so it is not a decision taken lightly. Economists talk about the 'price elasticity' of supply. Suppliers with low fixed costs (facilities, debts, permanent employees) can react quickly to drops in prices, but oil producers have fairly high fixed cost – they rent or buy land, hire or purchase expensive equipment, take out loans to pay for capital expenses and have highly-skilled employees. Sometimes it is better to continue production at little or no profit than run up (greater) debts or lose leases. To minimise losses in 2014-2016, producers stored oil and planned to sell it only when prices rose.

Affordability – the cost of extraction

In "What has the oil industry ever done for us?" you saw the remarkable, innovative, hard work that goes into fuelling our lives. As conventional oil production grew to a peak, prices started to rise and these innovations meant we could fill the supply gap with deep-sea, heavy and tight oil. The downside of these new (ish) oils is their high extraction and refining costs. It seemed we were in a cycle where rising prices justified more expensive extraction, and more expensive extraction kept prices high. In theory, this could lead to spiralling prices. However, the price drop of 2014-2016 put paid to any notion that oil prices could be predicted so easily.

Until that price drop, trying to get hold of publicly-available cost data was a thankless task. Then Citigroup and BusinessInsider (respectively, an investment group and business/ technology news site) shared previously-private data. Their chart (see the Sources) shows the breakeven costs for new projects planned up to 2020. For every dollar drop in price, more projects turn from profitable to unprofitable. The only reason for continuing with these projects is a belief that prices will recover enough to make them viable.

---

165 Really – price variation has an impact on price! Don't be so cynical; read on.

Affordability – economics

To be able to afford oil, the growth in our income has to keep up with the growth in its cost. If the price of oil goes up faster than incomes, we

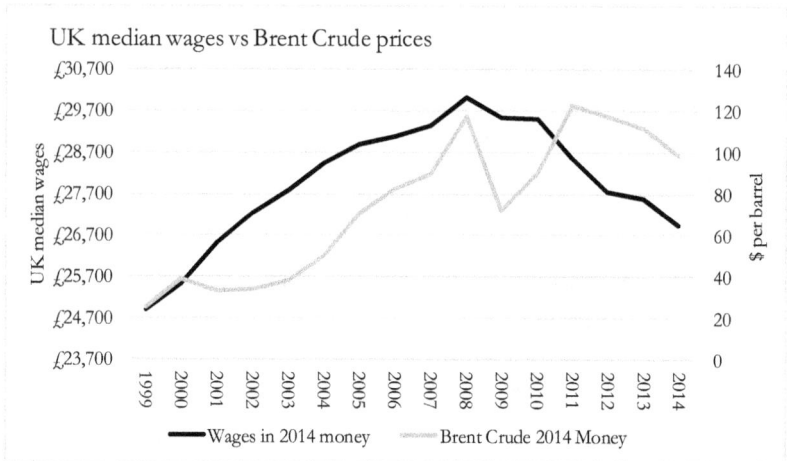

UK median wages vs Brent Crude prices

Figure 14: UK wages have not kept up with Brent Crude, even as it fell

can't afford as much of it. There are two ways of looking at this, and the first is to compare oil prices with median salaries. The chart below shows a few interesting things.[166] But mainly it shows that oil grew relatively cheaper until sometime in 2009, and then became progressively less affordable until it dropped in price. It didn't return to 1999 levels of affordability until mid-2015.

The second way is less obvious but equally important and hinges around inflation. Now, we are not going to get into the economics of inflation here – you can find many good books on that subject. Let's just cover the basics. If the pound in your pocket buys less this year than it

---

166 Okay, some people don't find them interesting – so here they are, relegated to a footnote, just in case you are a geek like me. Wages went up much faster than oil prices in the early Noughties, then, just before the 2008 crash, oil prices hit a peak, dropping as the world economy shrank. For a brief spell in 2009 it looked like both oil prices and wages would recover. But, while the oil price soared upwards, we got poorer. Then oil prices started to fall – and fall. Even more geeky: If you check the ONS website you will find they show that wages went up, because they used the 2015 median wage and CPI rate (both of which were provisional at the time of writing) and I used the 2014 figures.

bought last year (be it bread, paper, beer or petrol), then inflation has made your pound less valuable.

⊺ In 1989 one hundred of your fine British pounds would have bought **five and a half** barrels of oil.

⊺ But inflation decreased the value of those coins and, if the price of oil had risen at the average rate of inflation, by 2014 you wouldn't have quite been able to afford **three barrels**.

⊺ However, inflation is an average; the prices of some things rise more slowly than inflation. And between 2003 and 2014, oil prices rose considerably more quickly. As a result, in 2014 your hundred pounds only bought **one barrel**.

These charts show how oil prices have fared in comparison to inflation (CPI) since 1989.

Brent prices actual vs inflation

WTI prices actual vs inflation

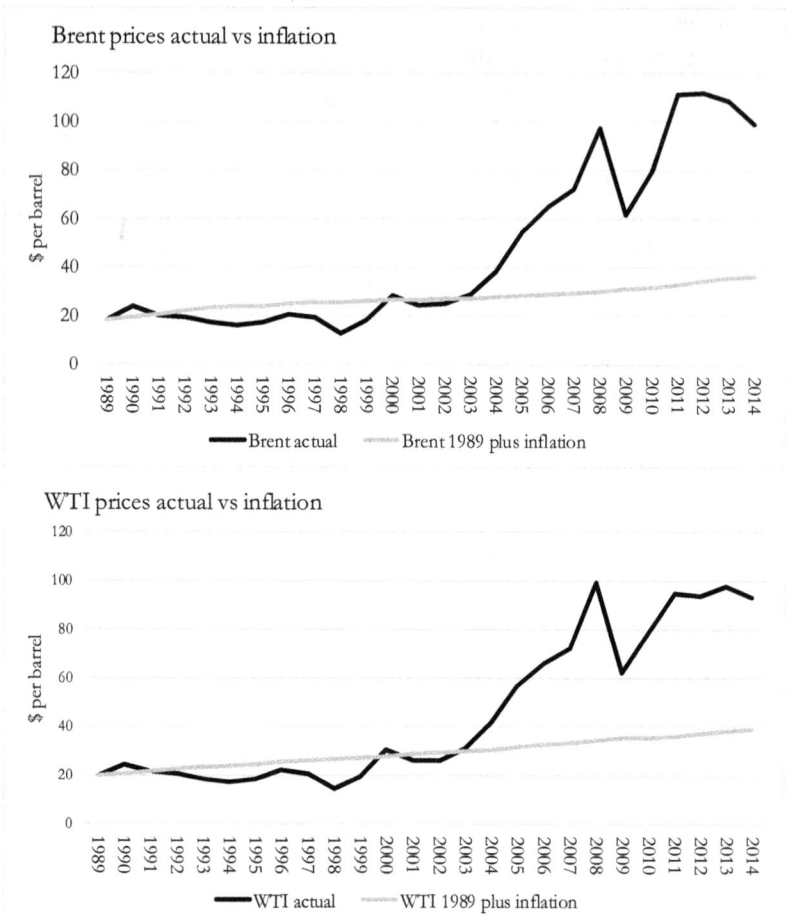

Figure 15: Oil prices rose faster than inflation from 2004

You'll remember that economists tell us the economy is growing, and that we are getting richer when GDP grows. So in addition to considering the recent direct relationship between oil, wages and inflation, we can ask whether the economy grows in a dependable fashion. If it does, the next chart will show GDP growth as a nice steady line.

Alas, it shows growth coming in lumpy, bumpy, unpredictable, undependable fits and starts. Growth slows regularly and even reverses (ie the economy shrinks), by turns driving oil (and other things, for that matter) into and out of our financial reach.

UK GDP growth

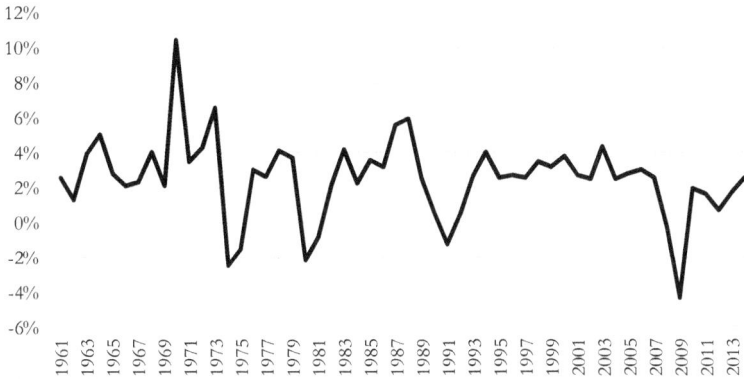

Figure 16: UK GDP

Affordability – political price pressures

If we wander too far into politics we may never escape so we will avoid that quagmire. The following are political subjects that influence the price of oil. You can find whole books on each of the following.

Energy security – 'let's produce our own oil and sell the surplus' – or at the very least, 'let's buy less oil from other countries'.

Economic strength – 'we have shedloads of oil; let's use it to keep our economy strong'.

Environment – 'oil pollutes our land, air and seas, and produces greenhouse gases'.

Human behaviour – 'I like money, power, comfort and all that oil brings me – I will do what I must to keep them'.

Financial markets –a frightening game of *The Price is Right*, which pushes prices up and down based on interpretation of events and 'sentiment'.

Demand – economic development

The industrially developing countries aren't going to stand by and watch us enjoy our lifestyles while they struggle to feed their children. They will use more oil as they adopt the relatively easy lifestyles of the richer countries. They will consume oil in farming by making irrigation systems, fertilisers and pesticides. They will consume oil to keep local industries in business and raise tax revenue. They will consume oil to make plastics and chemicals, and virtually every manufactured item. And there will be a strong demand for all these things for some time to come. For example, in the three years 2011 to 2013, China used more cement (6.4 gigatons) than the United States employed in the whole of the 20th century (4.5 gigatons). As you saw in "Oiling the wheels of industry", that's a lot of oil.

Global oil consumption

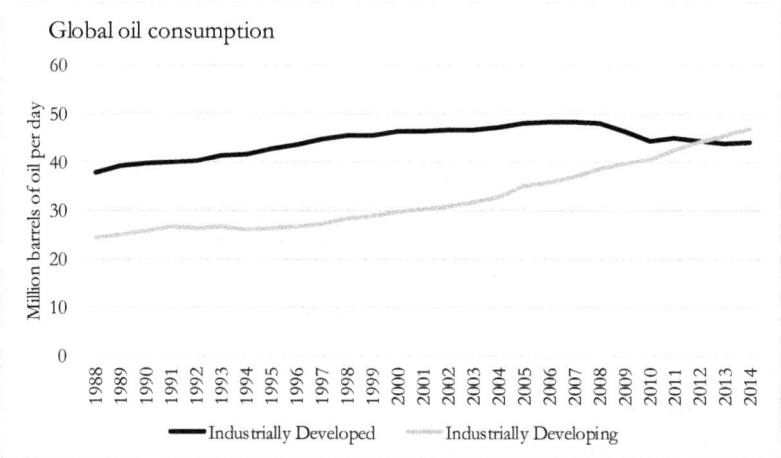

Figure 17: Global oil consumption

We can already see this happening in the chart above. The line showing the oil consumption of the industrially developed countries covers a mere 29 nation states. The other line – the one shooting upwards – shows the growth of oil use in the other (roughly) 160 countries. If those countries 'achieve' the same per-person use as the rich 29, they will use 16 times as much oil.[167]

---

167 There is a teeny, tiny problem with the data. Well, two, actually. First, some of the rich countries don't use much oil, so they are lumped into 'others' by the EIA. This makes a small difference to both lines. The other problem is that politics leads the

All in all, our best guess is that oil prices will remain volatile, fickle, unpredictable and undependable.

"Prediction is very difficult, especially about the future."

– *many writers from Mark Twain to Niels Bohr, Nobel Laureate in Physics*

---

people producing the data to carve up the world in different ways. This is particularly apparent with Chinese data: the EIA classes Taiwan and Hong Kong as separate countries (oil data) and the World Bank considers them to be part of China (population data). I haven't adjusted for this.

Chapter twelve

Peak oil

DEPENDING ON WHAT YOU READ, you might think that either 'peak oil' is a mistaken belief that fossil fuels will run out, or it is a forerunner of the end of the world as we know it. The truth lies somewhere in between. It is a fact that oil production will peak[168] and, by definition, supply will slow down. It follows that if demand does not match supply, prices will increase. Therefore, the economic impact of the peak will be felt far earlier than the end of oil production.

M King Hubbert built a mathematical model and worked out that American oil production would peak between 1967 and 1970. He presented his analysis to the American Petroleum Institute conference on 8 March 1956. They dismissed his conclusions because previously-predicted peaks had failed to materialise. Hubbert's model proved accurate in 1970, and a little more than 30 years later, American oil imports were almost three times domestic production. Hubbert continued to develop and refine his model, and in 1974 he predicted *global* oil production would peak in 1995, at 40bn barrels a year. However, we have never been that oil thirsty and the peak in the oil familiar to Hubbert happened in 2006.

---

168 Scarily, 'peak' can be applied to the production of many things – from the fossil fuels, to water and minerals.

Why was Hubbert's prediction wrong? Well, he did not expect the economic slowdowns of each subsequent decade, efforts to reduce carbon dioxide levels, nor the ascendance of natural gas. He did not foresee improved extraction methods including horizontal drilling, deep sea drilling and fracking. His model did not account for heavy oils. Economists argue that the errors in his second prediction demonstrate we will continually postpone the peak because we will develop new technologies such as those listed above. Indeed, technical advances have since lit a new dawn for American oil, and in 2014 imports and domestic production reached parity.

However, advances in science and technology cannot physically increase the amount of oil waiting for us. Indeed, with science and technology, we increase use. Nor can we pay for new technologies unless the price of oil is sufficiently high.

In 2005 William J Cummings told the Boston Globe that producing oil could only become more difficult (and, therefore, more expensive) and, in the same year, the United States Department of Energy commissioned Robert Hirsch to research oil availability and the likely effects of peak oil on the American economy and society. The introduction to the Hirsch report reads:

> *"The peaking of world oil production presents the United States and the world with an unprecedented risk management problem. As peaking is approached, liquid fuel prices and price volatility will increase dramatically, and, without timely mitigation, the economic, social, and political costs will be unprecedented. Viable mitigation options exist on both the supply and demand sides, but to have substantial impact, they must be initiated more than a decade in advance of peaking."*

> *– Provided by the US Department of Energy's National Energy Technology Laboratory*

A couple of years later the International Energy Agency stated that conventional oil production had peaked in 2006. Without US tight oil, the world would have produced only as much oil in 2013 as it did in 2005. That would have really pushed up prices.

Historic oil prices

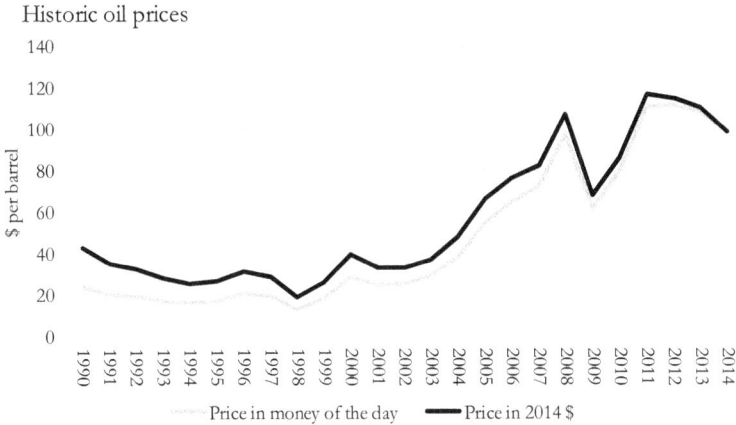

Figure 18: Historic oil prices

In fact, it did. Tight oil[169] filled the gap in conventional production, but it costs more to extract, and the extra expense of the tight oil industry accounts for some of the price rises in the chart above. At the point of the last peak, we were lucky that fracking technologies were waiting in the wings. Unless we prepare for the next peak[170] the results may be less serendipitous. As Dr Birol told the Independent in 2009:

> *"we have to leave oil before oil leaves us, and we have to prepare ourselves for that day."*
>
> – *Dr Fatih Birol, Director, International Energy Agency*

---

169 And, to be precise, other unconventional oils such as heavy oil from Venezuela and bitumen from Canada.

170 Did I mention that the US Energy Information Agency think US oil production will peak in 2020? At that point the real impact of the oil price crash will become apparent, as the oil companies will not have enough 'new oil' to fill the gap.

Chapter (lucky) thirteen

What's the worst that could happen?

REMEMBER ALL THOSE USES OF OIL we looked at in Part One? When oil prices are high, the impacts are quickly felt and easily understood. No matter how you use oil, high oil prices hit us and everyone we know in the pocket. They compromise our ability to do the things we enjoy and even stop many of us having access to essentials such as heating and food.

You might think 'that's alright – what goes around, comes around'; so when prices drop, we benefit. But do we? We will see shortly how little of the drop in oil prices in 2014-2016 was passed on to us. Worse: it had other impacts that were only felt immediately by those in the industry. It takes time for the consequences of job losses and lower tax receipts to hit the rest of us. The results of low oil prices are harder to understand than the immediate and obvious effects of high prices. These effects and their hidden, insidious nature mean they can catch us unawares (like bananas in pyjamas).

The downsides of low oil prices

In "It's the economy, stupid" we learned about the number of British people employed directly and indirectly by our oil and gas industry. Starting in 2014 and carrying on throughout 2015 and into 2016, these companies announced redundancies; they cancelled the contracts of self-employed specialists; and they negotiated lower prices from their supply chain which, in turn, also shrank operations. In November 2015 a recruitment specialist estimated that quarter of a million jobs had been lost in the global oil and gas industry. Tens of thousands of British jobs will be lost by the time prices recover; that's tens of thousands of families claiming benefits, tightening belts and no longer paying taxes.

> *The impacts of low oil prices: reduced company revenues, job losses, pay freezes, reduced income tax take, reduced expenditure in the economy and increased welfare bill.*

British oil and gas companies have enormous pools of talented people and knowledge. As the North Sea struggles, these companies become more vulnerable to takeover by foreign firms.

> *The impacts of low oil prices: skills and knowledge lost from the UK industry, taxable income lost to the Exchequer.*

We already know that our savings and pensions are dependent on oil prices. A Canadian research firm has calculated that a collection of pension plans worth $1 trillion lost $22bn when the shares of oil and gas companies fell with the price of oil.

> *The impact of low oil prices: damage to your pension, investments and savings.*

Since the UK started exploiting the North Sea reserves, the oil and gas industry has paid £316bn in production taxes. In the fnancial year 2014-2015 the public sector deficit averaged £3.8bn a month and the national debt averaged £1,776 billion.

> *The impact of low oil prices: reduced corporate tax take, increased public sector deficit and increased national debt.*

The UK oil and gas industry improves our balance of payments by £50bn a year. This comes from reducing imports and creating exports.[171] In 2013 our balance of payments deficit (aka trade deficit) was £123.7bn. Without the UK oil and gas industry this figure would grow by 40%.

---

171 In addition to manufactured goods, designs and services, we export about half of the natural gas we produce and around three-quarters of the oil.

The impact of low oil prices: increased UK balance of payments deficit, which weakens the pound. A weak pound improves our exports but makes imports (and overseas holidays) more expensive.

Extraction rates in the North Sea have been in decline for some time. This is because we have used up the oil in the easy-to-extract reserves and now the industry has to look to resources that are more expensive to unearth. This decline is despite the £14+ billion the oil and gas industry invests every year – the largest investment in any industry in the UK.

UK oil and petroleum products

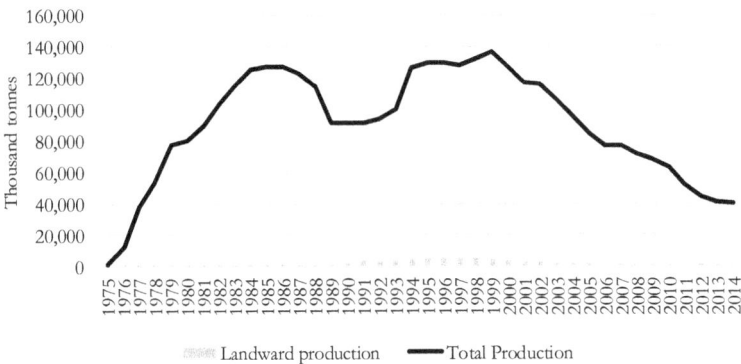

Figure 19: UK oil and petroleum products – production since 1975

There is plenty of oil under the North Sea (between 15 and 16.5 billion barrels[172]), and technically it can be extracted. But under the UK tax regime, it is not profitable. So, for a long time, the industry has been calling for changes to various taxes, and for financial support for development. When oil prices dropped in 2014, the industry raised its voice, and the Chancellor obliged with tax cuts in the Autumn Statement to stimulate investment and exploration. He delivered more in March 2015 though the emergency budget in July 2015 offered little extra support. Of course, the current trend in oil prices means these calls will be repeated for some time.

---

172 If you have remembered your stuff, you will realise that about two thirds of this figure are proven reserves and the rest unproven.

> *The impact of low oil prices (and peaking oil production): reduced corporate tax take and increased public expenditure (subsidies).*

We already know that shale operators continue to produce oil, even if they are losing money. By way of a refresher, here are the reasons: First, debt means that any income (or increase in the value of assets) is better than none. Second, in some fields, when operators stop producing they lose their lease (and any investment in the well). Third, production from wells in the shale plays drops quickly.[173] It takes a great deal of work (and money) to reopen a shut well, and short-lived tight oil wells may not justify the expense.

> *The impact of low oil prices: low prices may still not result in reduced production, meaning prices are likely to go lower still.*

When operators are losing money on each barrel they produce, clearly they hope for a price rise. To minimise their losses, they put away oil for a sunny day. The problem in knowing how much they are storing is that measuring stored oil is not like counting the tins of beans in your kitchen cupboard. It involves words like *contango* and *backwardation*, and specialists report oil stocks by the amount volumes changed, rather than the actual levels. In this case data lets us down, and we need to look in another direction to see whether storage is increasing. In February and March 2015 US press organisations on both sides of the political divide ran stories about oil storage. While they made different interpretations of the data, they were both clear that oil companies were storing oil in preference to selling it. (Of course, the saturated market forced their hand.)[174]

> *The impact of low oil prices: putting oil in storage is banking on higher prices, but it could equally result in lower prices (as it causes a glut in the future) and more bankruptcies.*

No one would need an exploration licence to find examples of companies slashing investment budgets in 2015. BP announced a cut of 20% in capital expenditure in February of that year. Chevron planned to spend 13% less on capital and exploration in 2015 compared with 2014.

---

173 The EIA states that 50% of tight oil wells' estimated ultimate recovery (EUR) takes place in the first three years.

174 In November 2015 an analyst calculated that the oil in storage would reach from Edinburgh to London if held in supertankers.

Shell announced reductions in organic capital expenditure by more than 15% in their year-end presentation in January 2015. By November that year, industry cuts in investment totalled £200bn.

Nor does 'lack of investment' have to be the billions of dollars of investment cut by the big oil companies – it can come in the form of reduced numbers of rigs, or even the cannibalisation of rigs that are out of use.

The impact of low oil prices: less investment in new wells slows future production rates and results in higher prices. It also slows the development of new technologies. Ironically, lack of investment during lean years leads some people to believe low oil prices are one of the best ways of guaranteeing high oil prices, as companies will face even higher investment costs just to maintain production levels.

We learned in "Peak oil" that US oil imports rose from 1970 and steadied with the advent of tight oil. These charts show the change in US import sources from November 1999 to August 2015. This date range was selected as the *total* amount of oil imported was more or less the same in both years.[175]

---

175 In case you are wondering about the tight-oil claim that the new oil supply reduced American imports: this definitely happened, if you compare today's imports with their peak around 2004 to 2007 (and the inevitable increases that were to come). If you look at the data, you can see current imports are higher than they were when fracking took off around 1997.

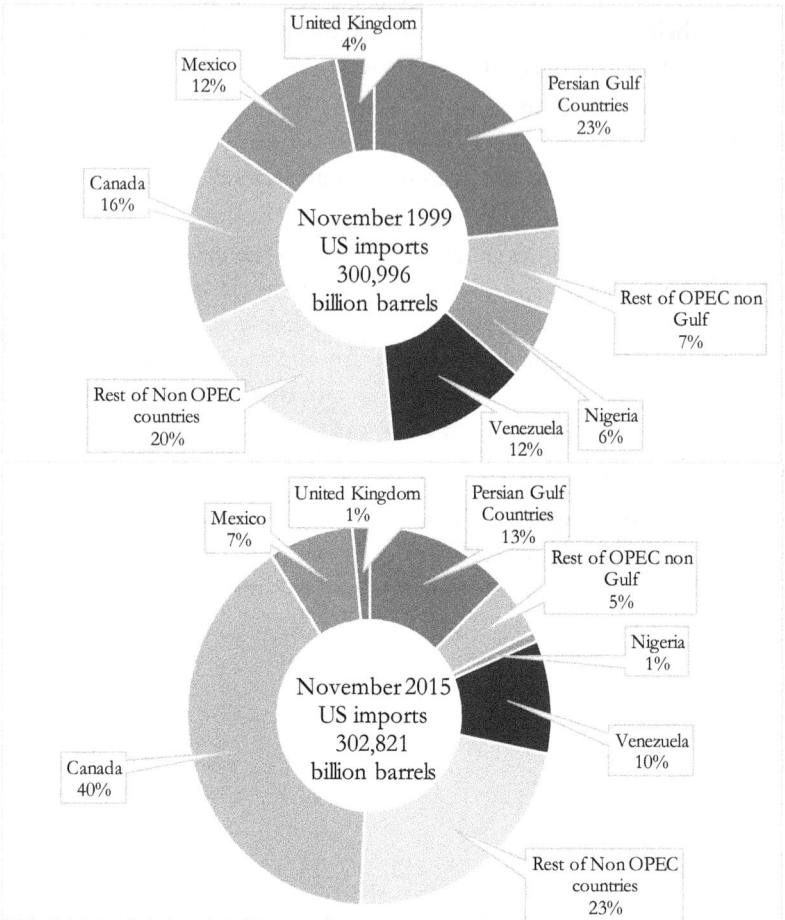

Figure 20: Changes in US oil imports

This change wasn't good for the countries that are heavily dependent on oil sales; for example, it resulted in budget cuts in Nigeria and a significant shrinkage of the Venezuelan economy. As the oil price slid downwards, the people of these countries – and all the countries relying on oil revenues – faced increased social and financial hardships and imported fewer material goods. And more people migrated to richer countries.

> *The impact of low oil prices: local economic woes. Even if you don't care about the people (don't mention the wars!... And migration), at*

least, note that such instability creates uncertainty in the supply (and price) of oil.

The 1980s were not just the decade of big hair, shoulder pads and great music. In the UK we suffered the impact of high-interest rates. In that

UK interest rates in the 'eighties'

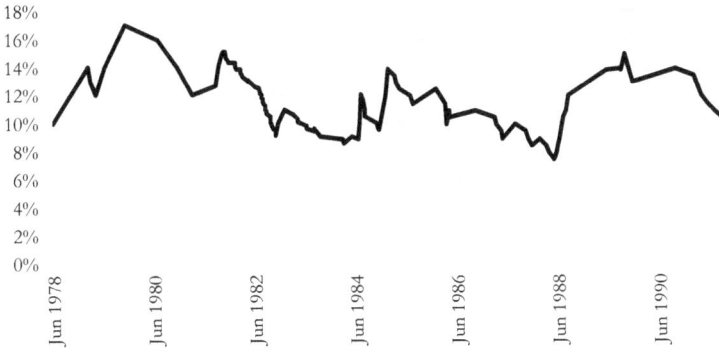

Figure 21: UK interest rates in the 'Eighties'

(approximate) decade the highest our interest rates reached was 17%.[176]

Russia is the largest European producer of oil (and natural gas), and one of the three largest producers in the world. In January 2015, low oil prices sent the rouble into free fall, and the Russian central bank responded with an interest rate hike of 6.5 percentage points, to 17%. They shaved two percentage points off in February, but that still left rates well above those we suffered through in the Eighties. Again, you may not care about the individuals, so imagine the actions that the Russian government might take to raise spirits at home.

The impact of low oil prices: devalued currencies lead to high-interest rates, which reduce investment in oil and cause global instability.

Who gains from low oil prices?

You might argue that the upsides of low oil prices outweigh these disadvantages. Let's check what benefits you are likely to have seen. A

176 This was on 5 November 1979, during the second oil shock.

lot of the oil we use has been transformed beyond recognition (for example, makeup or paint), but petrol and heating oil still bear some

How the price of road fuel (UK and US) responded to the drop in oil prices

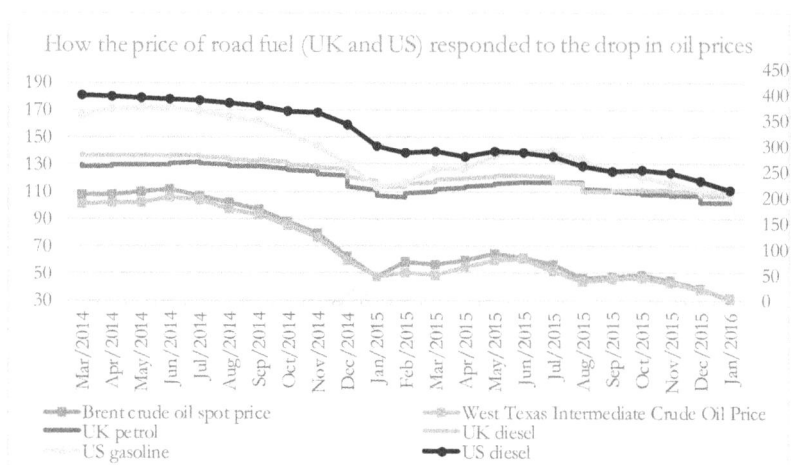

Figure 22: How the price of road fuel responded to the drop in oil prices

resemblance to the stuff that comes out of the ground.

Did their prices fall in line with oil prices? The chart above shows the price of road fuels and the price of oil in the UK and the US. You can see that UK petrol and diesel prices did not fall with the price of oil, whereas US prices did. (Don't worry too much about the actual numbers; just note the shape of the lines.)

Some people caution that UK petrol prices 'cannot' drop with oil prices, because of fuel duty and VAT.177 Of course, that would explain the situation – so let's see how true it is. Here comes the maths.

First we will make some assumptions about the components of the price of a litre of unleaded petrol:

- 𝕚 Fixed cost – fuel duty (58p).
- 𝕚 Fixed cost – retailers' markup (6.43p; remember, the retailer has to find all their costs out of this).
- 𝕚 The price of oil (this includes exploration, transport, production and refining costs as well as tax and profit).

---

177 In 2014 a House of Commons paper cited changes in the UK-to-US exchange rate as the second reason for oil price drops not having been passed on (in 2008). I will let you do the maths on that one.

⚡ VAT (charged at 20% of the total of the other costs).

Using the formula below, we can calculate the price of oil in a litre of petrol when petrol cost £1.29/litre in June 2014, and then work out how much it should have cost as the price of oil fell.

$$Fixed\ costs + oil\ price + \frac{fixed\ costs + oil\ price}{5} = price\ at\ the\ pump$$

Figure 23: The equation for the price of petrol

You can see the answers in the chart below. The bar shows the make-up of the price of petrol in June 2014. The solid line shows how the price of petrol changed. The dotted line shows how it could have changed. We could have been paying less than a pound per litre from August 2015, but these prices did not arrive until Asda 'slashed' their price on 27 November 2015. In the meantime, someone be it the oil companies, refineries or retailers benefited from the difference between the potential and actual prices (the exchequer definitely did).

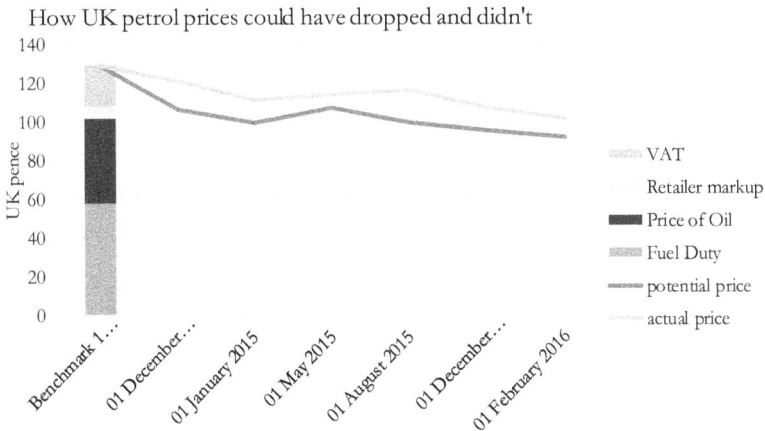

Figure 24: Make-up of the UK price of unleaded petrol, and changes in price

So much for road fuel. What if you live in one of the homes that contributed to the need to burn 16m barrels of oil in 2014? The graph below shows that neither of the two types of domestic heating fuel (aka standard grade burning oil and gas oil) dropped as far in price as crude oil did.

In summary: high oil prices are bad because we can't afford stuff. Low oil prices are bad because they damage the economy, don't get passed on, and (if you recall) didn't fall as far as the median wage this time around.

The price of domestic heating oil vs Brent Crude

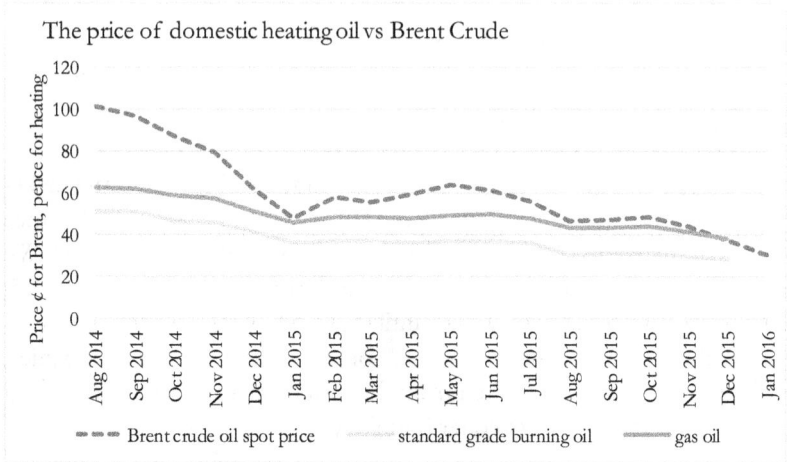

Figure 25: Heating fuel prices didn't drop with the price of oil

We need to ask if we are insulated against the 'human' downsides of using oil. History tells us that when people see an opportunity to make money, they may take selfish actions. Imagine businesses and governments getting access to oil. What practices and shortcuts might people be willing to take to get themselves a (larger) slice of the cake? Will our dependency on oil increase our readiness to accept drastic production methods, like strip mining, or to relax fracking controls? Will we tolerate, or even justify, oil spills in the race to satisfy demand? Will energy-poor communities turn to crime or terrorism to survive? Will we act to avoid this fate?

When we see the impact of volatile, fickle, unpredictable, undependable oil prices, we understand the doomsayers – but generally, we choose to ignore them. We deny the need, we discredit the alternatives, we argue that change will be worse than staying the same. Now that you see what could happen, if we continue to depend on undependable oil, are you more likely to do something?

Other impacts of depending on oil – Nigeria as case study

Some of the people who have studied the subject of oil prices predict that high prices will tear apart communities, and irresponsible extraction will spoil our planet. They predict social breakdown and public unrest on a scale never before seen. Low prices draw as many woes: the oil industry hunkers down and spends less money, every part of the supply chain feels the hurt, and governments get less oil revenue. Nigeria provides a cautionary tale for both high and low prices.

Nigeria began extracting oil in 1958. In the Seventies, the country soon became one of the largest oil-producing countries, providing over 7% of American oil imports. You might expect the country to have benefited from oil revenues. Yet the Nigerian people lost land and access to food; they remained poor. The Niger Delta fell into revolt (or protest, depending on your point of view) and a government-led amnesty failed to stop violence and sabotage. Now, oil thefts abound; the oil companies blame the thieves for most oil leaks. Gas flaring persists because burning gas costs less than taking it to market or giving the people of Nigeria a small bonus. Today, half the Nigerian population has no access to electricity.

Human Rights Watch regularly reports on Nigeria and in 1999 related a roster of statistics about oil spills. They obtained government records that showed almost 5,000 reported spills from 1976 to 1996. The spills released nearly 2.5m barrels. The same data estimates that the best part of 2m barrels were 'lost to the environment', which means they still lie on land or in lakes and rivers. In 2011 the United Nations Environment Programme calculated that Nigeria would spend more than 30 years cleaning up the pollution (that's 30 years from when the clean-up starts, not from the date of the report). During each year of extraction, more oil is spilt in Nigeria than polluted the Gulf of Mexico in 2010. Unlike the BP spill, holding the oil companies to account is not an inevitable fact of life. In January 2015 Shell agreed to pay an out-of-court settlement of £55m to individuals and the community of Bodo in the Niger Delta. While this is possibly the largest settlement made in Africa, it pales into insignificance compared with the BP court-ordered settlement of $46bn for the Gulf of Mexico spill. What's more, it took over three years to achieve the

Bodo deal, which was a significant improvement on the initial offer of £4,000.

Then came America's tight oil boom. A desire to reduce dependence on oil from the Gulf states provided a compelling argument to see fracking as a patriotic act. As we saw earlier, Nigerian imports fell off a cliff.

Nigeria tried to sell into the Asian markets, with limited success. Then the price of oil plummeted. Nigeria set about a new budgeting round, raised taxes and decreased government expenditure. The country borrowed heavily to make up shortfalls in income and now faces years of debt and austerity.

You might argue that other countries have stronger controls and would not repeat Nigeria's experiences. Yet no oil-related problem occurs solely in the Niger Delta. For example, Nigeria has no monopoly on the economic damage caused by lack of development and growth in some areas when wealth rapidly increases in another (for example, from oil exports). Indeed, the phenomenon is called the 'Dutch Disease', as this happened in the Netherlands over 60 years ago when the oil wealth of the North Sea was first exploited.

We also call these effect the Oil Curse. Rising food prices in 2008 provide a global example of the curse. The increasing cost of food caused food riots, protests and government interventions in many countries; eg in Panama, where the government subsidised rice prices.

The World Bank took some months to say that biofuel production (alcohol from maize) accounted for 70 to 75% of the food price increase. Three years later, the International Food Research Agency reported on the causes of the 2005-2008 food crisis, which they compared to the 1973-1974 food crisis. They found a near-perfect storm that led to the later event. Causes included increasing energy prices, growing demand for biofuels, depreciation of the American dollar, various trade shocks triggered by export controls and panic buying, and bad weather. The timings show the meticulous nature of the assessment; the analysts used widely-accepted data to develop their conclusions, which point (in different but undeniable ways) at undependable oil prices.

Chapter fourteen

Where has all the oil gone?

OUR EXTRAVAGANT USE AND NEGLECT of oil litter the land, sea and air. So, any consideration of the price of oil must include disposal costs. On this subject, we need to turn to environmentalists for several reasons:

- Their voice ensures data about oil use is available publicly. Using carbon dioxide to estimate oil use may be imperfect, but is better than having no data.

- They have shown us that we have wasted, and continue to waste, valuable resources through pollution.

- They started the journey to a world less dependent on oil. Without them, the alternatives we have considered would not be as ready as they are.

- They have won protection for the environment and press for more.

- They have demonstrated that we cannot afford to burn much more oil.[178]

---

178 The theory of unburnable oil has been around for a long time. Simply put, it says that to avoid irreparable damage to the environment we can burn no more than 20% of known fossil-fuel reserves (this number changes with time, discoveries and our growing understanding of climate change). Financially this gives us a huge problem: the value of

In this chapter we will consider the ways in which oil pollutes the land, sea and air, and then turn to climate change. Of course, we end with a look at the many people who are reducing the impact of oil on our health and our planet. In between, we will look at the environmental arguments in the debate about fracking.

Land

The act of extracting oil is brutal. Operators tear up vegetation and remove fertile soil, and then they add a liberal coat of concrete. Wells drive into the ground, churning up rocks and clays, which form heaps or fill dumps. A crisscross of roads and pipelines sprouts up like bindweed. Noise is ever-present, and lights mark each site throughout the night.

Countries with sizable oil reserves have lost forests and pristine ecosystems. For example, Australia enjoys greater hydrocarbon riches than most countries. In addition to vast reserves of coal, the Land Down Under sits on a wealth of natural gas and oil. Exploitation of these reserves comes at the expense of farmland and spectacular areas of natural beauty and biodiversity, such as the Kimberley region.[179]

What's more, the impact of extraction goes deeper than you might think. Oil, gas, coal and water hold up the land, and in the wrong geology extraction causes subsidence. Large tracts of land in Indonesia are sinking because of the extraction of groundwater and natural gas. Nor is this archipelago nation alone; geologists first noted subsidence caused by hydrocarbon extraction in Texas early in the 20th century and have since documented the same effect in many other countries.

At the other end of the supply chain, sturdy plastic bins welcome our refuse. Lorries truck waste from our homes and businesses and commit it to a dank and smelly grave. In short, we dig holes and fill them with

---

companies and the economy of many countries is based on the amount of oil we are able to extract (and sell). If we can only burn 20% of stated reserves, then it follows that some companies must be overvalued and national budgets that rely on oil revenues cannot be balanced. This risk is sufficiently significant that the Bank of England is investigating and the Financial Policy Committee reviews the subject regularly.

179 Even oil resources are fickle. Australia has very light oil. It is so light that heavy oil imports are increasing to complement their domestic product.

oil. A lot of oil.[180] The waste going to landfill includes plastics; cans; card and paper; compost; fridges and other domestic appliances; textiles; vehicles; electrical goods; chemicals; and batteries. In 2014, England's household waste weighed nearly 27m tonnes.

When it comes to packaging, the UK government reports that we recycled 65% and recovered 73%[181] of our commercial and domestic packaging waste in 2013. These levels were a big increase on 1998 when we recovered a paltry 27%. However, our efforts are heavily weighted towards card and paper; we sent 69% of our packaging plastics to landfill.

Landfill and waste recycling centres are unpopular for many reasons. We do not want them near us, yet we are all willing to continue transporting our oily waste to someone else's doorstep. And we are so keen to keep our wasteful habits hidden that we export the most controversial items. For example, it is illegal to put electronic waste in landfill in Europe. We do not recycle much electronic waste at home, so we ship it to industrially developing countries, with the vague hope they will. But recycling doesn't always happen; instead, the waste piles up in dumps where people, including children, scavenge the valuable materials from the poisonous without training, tools, or safety equipment. Although most governments have pledged to stop the irresponsible management of waste, we have some way to go.

And it's not only overseas that we dump irresponsibly. Some people take landfill fees as an excuse to throw their rubbish anywhere. In England in the 2014-2015 financial year, local authorities cleaned more than 900,000 incidents of fly-tipping. They spent nearly £68m on cleaning and enforcement actions in the same year. Private landowners cleaned up an unknown amount, at an unknown cost.

---

180 We will cover the limits of the following data in "To oil, or not to oil?". In the financial year 2010-2011, each English person threw away 429kg of domestic waste. Councils recycled, composted or reused 42% – in other words, 249kg went to landfill. But we have got better: we only created 410kg in the financial year 2014-2015 and, in the same period, we increased the proportion of the waste we recycled, composted or reused to 45%. Yay for us.

181 Did you ask yourself why recovery is higher? It adds composting and energy production to the total.

When you read the list of oil spills earlier you probably recognised the sea-based spills more readily than those based on land. However, land sees more than its fair share. In America land-based spills increased by 18% from 2012 to 2013 and totalled 20 a day. In Russia and Nigeria, where reporting is less robust, spill information might be anecdotal, but it suggests large amounts of pollution. Land-based spills are just as dangerous as those at sea and just as hard to clean up. Indeed, land, water and communities that experience an oil spill rarely return to their former condition.[182]

Despite the severity of oil spills, people who seek to bypass legal disposal methods cause more pollution. Engine oil, oil from industry, excess fertilisers and other chemicals reach waterways and make their way to lakes and the sea.

Sea

Most of the oil in the sea does not come from spills. Depending on who you ask, the figures differ, but they tend to agree on the sources and the main culprit. Oil in the sea comes from natural seeps, discharges from ships not in the oil industry, land-based sources (eg engine oil), oil operations (including transport), and atmospheric hydrocarbons. In the studies you will find in the Sources, land-based sources always contribute the most to the oil-based pollution in our seas. Yet, we hear little of this, as the speed and concentration of spills make them more dramatic. Not included in these figures are the land-based chemicals we use to produce food.

Without fertilisers and pesticides we cannot produce enough food, but these chemicals run into waterways and groundwater. The effect of fertilisers on seawater runs to two extremes: excess growth, and death. In 2011, seaweed and algae swamped beaches in France; in 2013 a fluffy, green algae blanket covered 29,000km^2 of the Yellow Sea; and in 2015 Lake Erie suffered its most severe algal bloom on record. In each case,

---

182 Even planned restoration, where the top levels of soil and vegetation are removed and stored, does not return land to its former state. When oil spills, we get no warning and have no time to prepare. Mobile animals flee if they have somewhere to go; otherwise, all living things perish. Cleaning up areas polluted by oil does not restore habitats, nor the plants and animals that depend on them. When inhabited areas are polluted, lives are destroyed, homes and jobs are lost, children are sent to new schools and communities pass into history.

fertilisers running off the land stimulated a growing spree. Equally common are 'dead zones': algae and bacteria grow unchecked when excess nutrients (sewage, fertiliser or methane) pour into the sea. They use all the oxygen and everything dies. The effect is reversible; nevertheless, the world's area of dead zones is as large as New Zealand. Other oil products reaching the sea are industrial chemicals, toiletries, sun lotions, medicines and plastics.

Some plastic stays in use for a long time, the rest we discard. We send most of our cast-offs to landfill, and litter with some of the rest – a small proportion that nevertheless has a high volume. Between 4.8 and 12.7 million tonnes a year makes its way to the sea and collects in the North Pacific Gyre. This strange beast is not a cousin of the Jabberwock; it is a vortex created by currents and winds spun by the Coriolis effect. The gyre lies thousands of miles from land, yet its waters testify to the care we take for the environment. They brim with small pieces of plastic washed into the largest waste site on Earth. At the surface each square kilometre holds 334,271 pieces.

Scientists call the rubbish collecting in the North Pacific the 'Great Ocean Garbage Patch' and have known about it for decades, but each ocean has a gyre and we now know plastics are gathering in all of them. Indeed, there are five trillion (that is, a five followed by 12 zeros) pieces of plastics floating in our oceans, and that could be as little as 5% of the plastic that gets there (the rest ends up in ice, at the bottom of the sea or on beaches). Nor must waters be salty to suffer from plastic pollution. Researchers chose Lake Garda, Italy to study plastic waste in freshwater, as they thought its remote location would mean it had little. However, the sediment at the bottom of the lake contains levels of plastic similar to those on marine beaches.

When plastic is subject to water and sunlight it splits into smaller pieces and individual molecules. It absorbs DDT and releases PCBs (polychlorinated biphenyls). Marine animals eat plastic items, which can kill them, and tiny fragments enter the food chain.[183] Under a microscope, you can see plastic particles in the digestive tracts of the crustaceans that feed fish and, ultimately, us.

---

183 90% of sea birds have eaten plastic.

We have already touched briefly on DDT. We used this pesticide globally to wipe out insects carrying diseases like typhus and malaria and to protect crops. Its disadvantage was the extinction or near-extinction of certain animal species. In the UK we lost most of our otters, and America nearly saw the complete loss of its national bird, the bald eagle, and the peregrine falcon. The chemical causes human diseases and many countries have outlawed its use. DDT's brothers in reputation are the PCBs. They are a family of chemicals with industrial applications from plasticisers in concrete to electronics. Like DDT, PCBs cause disease, and poison the food chain. Animal rescuers have found them in the organs of Californian sea lions with cancer. The photographs in the Sources might strengthen your resolve to waste less oil. If they do not, remember: sea lions eat the same fish as we do, and probably picked up lethal doses of chemicals from their food. All governments have banned PCBs, but they do not break down in the environment, so we're stuck with them.

Given a bag or bottle of these chemicals to eat or drink, the skull and crossbones would encourage you to say no thanks, but in reality, you swallow them all the same. These chemicals have two characteristics that amplify their effects as they move up the food chain. Animals cannot digest or expel the chemicals, so they hang around in their (and our) bodies; this is known as persistence. When a large animal eats many small animals, the concentration increases; it is called accumulation. Animals at the top of the food chain (large fish such as tuna, sea mammals and us) contain up to 10,000 times more DDT and PCB than the oceans on average.

In a neat turn of events, we have found microbes consuming hydrocarbons in the North Atlantic Gyre. Perhaps, one day, others will break down PCBs. No one knows how quickly the bacteria will gobble up ocean pollution (or if they are just speeding the entry of pollution into our food chain), but if we stop polluting today, we might bequeath clean seas to our distant descendants.[184]

However, the shocking pictures of pollution from ships that have reached the end of their useful lives may force us to reconsider that hope. Ocean-going vessels pull on shore in industrially developing countries.

---

184 It strikes me that these bacteria, should they evolve to create a society, could one day communicate about Peak Plastic and the need to learn how to live on land, where explorers have found huge food resources called 'Lanphil'…

Here, workers break up the ships and send most materials for recycling. But nothing can entirely stop the escape of pollutants. Chemicals leach out of rubber parts, toxic chemicals are used to clean tanks, then spill out, and oil runs from ruptured pipes. These chemicals pollute the beaches and sea.

And the oil industry has another negative watery effect. The sonic guns we learnt about earlier may deafen marine animals.[185]

Cleaning up oil has created an interesting spin-off industry. Traditionally, *booms* stop oil spreading, *skims* collect oil from the surface of the water and *detergents* dissolve the oil. These methods are themselves oil-hungry, though, and are not tremendously effective for large spills. Even worse, some of the chemicals used in clean-up slow the natural decomposition of oil by bacteria. However, scientists are rising to the challenge of improving cleaning technology. They are inventing methods that use less oil and are reasonably effective for small spills. If oil disperses before we can capture it, it will eventually decompose, into various chemicals including carbon dioxide.[186]

Carbon dioxide has two big effects on the seas (which we will review in the Climate Change section), but we emit it into the air.

Air

As we burn oil, we release several chemicals; the most familiar is carbon dioxide. We shall consider the effect of increased levels of atmospheric carbon dioxide in a second. First, we must remember that oil is not all hydrocarbon. All the fossil fuels contain other substances. In the last century, nitrogen and sulphur released from coal killed thousands of people through smog and destroyed forests with acid rain. Your combination boiler releases sulphur. Steam picks up the sulphur, condenses and flows into drains. There, it rots our, largely Victorian, sewers.

Volcanoes and burning hydrocarbons release benzene. Since the 19th century, this ring of carbon found various uses, including as aftershave.

---

185 Of course, oil extraction also uses a lot of water. We will see shortly that this is on a par with golf courses and clothing manufacture.

186 That 'eventually' covers a multitude of sins. Oil destroys life in the seas and lingers for decades – to be honest, we don't really know all the effects of oil spills, particularly in locations that have been lucky enough not to suffer many so far.

Which is unfortunate, as it is a human carcinogen, affects blood health, and harms foetal development. When we stopped adding lead to petrol, we substituted benzene. It is also a stepping-stone in making plastics (polystyrene, polycarbonate, epoxy resins and nylon). When we burn plastics at low temperatures, they release benzene, dioxins and other chemical nasties. Benzene is a prolific oil product and is in the air you breathe, the water you drink and the food you eat.

Dioxins are another group of chemicals that are persistent and that cause diseases. They come from many sources, from volcanoes to the manufacture of some weed killers and pesticides. They are also released by combustion (forest fires, domestic hearths, etc). Opponents to burning municipal waste often cite the release of dioxins as a reason to avoid this form of electricity generation. However, studies have shown that the amount of dioxins released from modern waste-to-energy plants is significantly less than ambient levels during standard operation.

Methane leaks into the air at every stage of its journey from the rocks beneath your feet to the hob heating your baked beans (it also leaks from natural seeps, coal seams and people who eat baked beans). Methane in the atmosphere is not good. It is poisonous and prone to explosion and is a pernicious greenhouse gas.

The impacts of waste and pollution

- In 2012, Dara and the Climate Vulnerability Forum calculated climate change and carbon pollution are already costing the world $1.2 trillion per year and caused 4,975,000 deaths in 2010. They believe the number of fatalities will rise to 5,957,000 per year in 2030.

- Fly-tipping in the UK cost £62 million of public money in 2013-2014. Just one private landowner, the Canals and Rivers Trust, reported 700 incidents in 2010, costing £300,000 to clean up.

- UK beach cleaning costs €18 million (£13.5m) a year. During a campaign to clean British beaches in September 2014, the Marine Conservation Society found 2,457 pieces of litter on every kilometre of beach.

- In 2012 the UK emitted 467 billion kilograms of pollution into the air.

⚥ In the UK in 2012, we issued 3.5 billion kilograms of pollutants into water. The result is that only 27% of the waterways in England support a viable ecosystem.

  ➢ *Managing the South West river basin costs over £60 million a year. A project to improve the quality of its rivers will cost an additional £3 million and yield benefits of £44 million.*

  ➢ *Ofwat expects the water sector to spend £44 million from 2015 to 2020 to improve water services and resilience, and to protect the environment.*

⚥ Urban clean-up: Keep Britain Tidy calculates we spend £1 billion of public money picking up litter in streets. As they say, we could spend the money in much better ways. In 2014 the EU calculated the annual cost of cleaning up litter in the entire EU region to be £10 billion.

⚥ Once ancient trees, balanced ecosystems and freshwater reserves are gone, they are gone for a long time. No amount of restoration can put them back in our lifetimes. The financial cost of these losses is comparatively small, but the effect of fewer green and wild spaces on our health is incalculable.

⚥ The University of Stuttgart estimated that the pollution from EU coal-fired power stations causes 22,000 premature deaths. Five million lost working days accompanied the fatalities.

Climate Change

The time has come to speak of climate change, by which we mean the impact of releasing greenhouse gases into the wild. But we are not going to cross-examine the evidence. We will accept the number of scientists who have published findings showing we are changing our climate as proof (if you are not sure about that, you can read more in the appendix about climate change). Instead we will content ourselves with an exploration of the ramifications so **you** can decide whether you want to run the risk.

For the sake of brevity, we will only summarise five negative physical effects of increasing greenhouse gas levels. Of course, you might ask whether it's all bad; after all, every cloud has a silver lining. There is one much-lauded benefit: some people believe farming will become much more productive.[187]

Sea level rises

When air temperatures rise, glaciers and ice sheets – such as you find in Greenland, at both poles and on the tops of mountains – melt. This newly-released water adds to sea levels. At the same time the temperature of the sea increases and the water expands. This expansion doubles sea level rises.

Every assessment made by the IPCC (see the appendix on climate change for an explanation) estimates the sea level rise we face. In 2013 they used several scenarios to explain the risk. For example, it was thought we could hold the average global temperature rise to 1°C by slowing emissions from 2010 to 2020, and then keeping them at this lower level. In that scenario the seas would rise 0.26 to 0.55 metres by 2100. Unfortunately, we are set to sail right past that temperature. The next scenario assumed that if we keep increasing emissions throughout this century, then the global-average temperature will rise by 2°C and the seas will rise 0.45 to 0.82 metres by 2100.[188] Estimates developed by other scientific groups increase the range in both directions.

Rising seas engulf land that lies by water. Many urban areas are close to the sea or rivers. At best, they will see their foreshores disappear. More realistically, many cities will experience severe flooding and storm damage.

A couple of other impacts of warming water

Glacial meltwater, on its way to the sea, floods the valleys and plains that get in its way. And many lakes are also warming up, but they tend to evaporate rather than rise.

---

187 One potentially positive outcome of increasing carbon dioxide levels in the atmosphere is an increase in the growth of some kinds of plants. However, the number of caveats and limitations of this benefit mean it cannot be given more space here – there is a link in the Sources, if you want to read more.

188 We will see shortly that sinking and rising land alters the relative sea levels in many locations. Here are some (even) less obvious reasons why some regions will see greater sea level rises than others. At a brisk trot: Melting land ice will change the gravitational field of the Earth. The weight of the additional water will squash soft sandy sea beds.

Changing weather patterns (extreme events)

Rising air and sea temperatures have changed weather patterns. Heatwaves in Australia, Asia and Europe are becoming more frequent, severe rainfall events are increasing, and East Africa is drying out more quickly.

Just a small increase in temperature can increase the number and severity of extreme weather phenomena. Until recently we have been unable to point to individual events and scientifically demonstrate they may have been caused or made more severe by climate change. However, you might have seen recent news stories showing how climate change increased the likelihood of flooding in the UK. If you want to read more, the Sources have a couple of useful links. Severe weather events include:

- Storms and tropical storms (hurricanes, typhoons, cyclones), eg the January to February 2014 winter storms;
- Heatwaves, eg the 2003 heatwave, which caused up to 2,000 additional deaths in the UK;
- Heavy and unseasonal rains, eg the Boscastle flash flood in August 2004.

(NB The examples illustrate event types from a UK perspective, with no suggestion climate change caused or worsened them.)

Ocean acidification

The seas are not only warmed by increasing carbon dioxide levels, but they also absorb the gas and become more acidic.[189] Both changes affect marine flora and fauna. Coral dies and takes thousands of species down with it. Jellyfish thrive and eat all the fish. Algal blooms increase. Whales starve on their annual migrations. And few changes to the ocean environment are good news for anyone depending on the seas for their living or food.

---

189 Acidity is measured on the pH scale, which runs from zero to 14. Anything with a pH less than 7 is acidic; substances with a higher pH are alkaline. Water is neutral, at 7. The sea is alkaline, but its pH is dropping (becoming more acidic). It is unlikely that the seas will become acid, but their pH has dropped by 0.1 in the last two centuries to 8.1 today. This doesn't sound like a lot, but represents a 25% increase in acidity.

Melting methane hydrates

The fossil fuels appendix describes a type of ice called methane hydrate. It lies below the seabed in many areas, and could become a source of natural gas. Low temperatures and high pressures keep the methane trapped in this icy form. As sea temperatures rise, the hydrate will break up and release its methane. Several things could happen next: Chemistry and microbes might convert the methane into carbon dioxide; those processes use up oxygen and suffocate ocean life. The sea could absorb the carbon dioxide and become more acidic, or the CO_2 could bubble up from the ocean and act as a greenhouse gas. Of course, it need not be that complicated; if there are not enough bacteria to 'eat' the methane, then it will escape into the atmosphere and perpetuate global warming.

Subsidence

The permafrost in Alaska, which provides a stable foundation for roads and buildings, is melting as the land warms. A quarter of it could be gone by 2100. While this example is not familiar, it deserves a quick mention to help us all remember that the impacts of a warming atmosphere may not be immediately obvious.

Economic impacts

The cost of our oil habit is heavy. We pay in the ways we've just seen and financially. Calculating the cost of climate change and pollution is a tough job, but someone chooses to do it. Here are some estimates of the cost:

- We have already seen the size of the costs of climate change and pollution as calculated by Dara and the Climate Vulnerability Monitor, so let's put the GDP figure into some context. It equals:
  - ➢ *2% of world GDP;*
  - ➢ *10% of the UK's National Balance Sheet (the wealth or net value of the country's assets);*
  - ➢ *Nearly two Apple Corporations;*
  - ➢ *About half the wealth of the 400 richest people in the US (2015) or one-sixth of the wealth of the world's billionaires;*
  - ➢ *The Gross National Income of the poorest 73 countries.*
- The World Health Organisation (WHO) estimates there will be an additional 250,000 deaths per year between 2030 and 2050 due to

heat-related mortality, coastal flood mortality, diarrhoeal disease, malaria, dengue and under-nutrition. They did not include the effects of economic damage, major heatwaves, river flooding, water scarcity, migration or conflict. This number comes from a consideration of what will happen if climate change does not get worse. The calculation makes the WHO figures look much lower than Dara's (which, as we saw a few pages ago, predicted nearly 6m additional deaths per year in 2030), but they are over and above that death toll.

☦ The analyst group Maplecroft believes 67 countries face 'high' or 'extreme' risks from climate change. In addition to the familiar concerns about storms and flooding, they draw attention to the number of days when temperatures exceed 25°C. On these days, called 'heat stress days', it is too hot (and often too humid) to work. By 2045 nine countries could have half their year taken up with such weather.

☦ In 2007 the former Secretary-General of the United Nations, Kofi Annan, founded a not-for-profit organisation to study and resolve humanitarian issues – the Global Humanitarian Forum. During its short life (it closed in 2010 due to a lack of funds) it examined the impacts of climate change on the lives of the most vulnerable people of the world. It found that the changes wrought by carbon emissions kill 300,000 people a year, affect 325 million people seriously, and have caused economic losses of US$125 billion. It estimated that 4bn people are vulnerable to these changes, and a further half a billion are at extreme risk.

☦ In March 2014 the IPCC Working Group II released its report *Climate Change 2014: Impacts, Adaptation, and Vulnerability*. In Europe alone it has observed the effects of climate change to date as being 'significant' in marine ecosystems, terrestrial ecosystems, human health, increased snow and increased wildfire. Each of these impacts has social and economic costs.

☦ A study in *Nature* concurs on the costs of climate change (it gives a 23% decrease in global incomes by 2100), but it also shows how the negative economic impacts are skewed towards hotter countries. Indeed, some colder countries may even see financial benefits.

꙰ Australia's devastating bushfire season in 2009 included Black Saturday. As the baldest statement of fact, the Royal Bushfire Commission estimated it cost AU$4.4 billion.[190]

꙰ Without changes to flood defences, the global cost of inundation could reach US$1 trillion per year by 2050.

꙰ Pollution and climate change cause death and ill health. The real cost of the death of loved ones is immeasurable. At the same time, to understand the effects of environmental changes on our economy, we need to put a price on death and illness.

> *In the United States, the Environmental Protection Agency (EPA) estimates health and mortality benefits arising from the Clean Air Act 1990. By 2010 the Act had led to 164,530 fewer deaths, 13 million fewer lost workdays and 3.2 million fewer lost school days.*

> *In the UK, neither NICE nor the Department of Health publicly put a price on our lives. But in the US, the American Cancer Society and Livestrong used data from the WHO to determine how deaths from cancer affected the economies of 188 countries. They worked out that the 7.6 million global cancer deaths in 2008 removed US$895 billion (1.5% of GDP) from the world's economy. This bill does not include any medical costs (which increase GDP).*

꙰ Mitigating the effects of climate change is also expensive. If we consider flooding, we traditionally respond to flood threats by building big walls, dikes and levees. However, times are a-changing and people are restoring rivers to their natural courses to prevent flood damage. There are also calls to replant upland areas (with trees) to slow the run of water into towns and cities. New and traditional approaches both come with the loss of land and significant costs.

꙰ The Environment Agency estimates that building and maintaining flood defences will cost £1 billion a year by 2035. While that is a lot of money, the annual cost of flooding is already £1.1 billion and may increase to £27 billion by 2080.

---

[190] Of course, a monetary value is the coarsest of expressions. The report detailing the fires is a humbling read.

Personal impacts

It is often hard to understand how the economics could affect you or me, so let's take it down to an individual level:

We know much of the UK is vulnerable to flooding, but probably do not realise the extent of land loss that will come with rising seawaters. We have learnt how much the IPCC WGII fifth assessment report (March 2014) expects sea levels to rise. Let's be honest; who can understand what a 0.45-metre rise in sea level will look like? The Environment Agency maps, linked in the Sources, will give you some idea of the implications for your home. If you want to go further, the United States charity the Natural Resources Defense Council uses maps to show the **predicted effect of climate change**. They show the extent of climate change impacts and cover of all the things we have been talking about and some more. Further still, you can look at National Geographic's maps, which show seawater levels if all the Earth's ice melted (an exaggerated, but interesting, position).

When you see how vulnerable (or safe) your home is, spare a thought for people living on **Pacific atolls**. Most of the atolls are four to six metres above sea level; however, much of Kiribati's land stands at only two metres. Tides already encroach into fresh water reserves and spoil agricultural land. The president of Kiribati (which is north-east of Australia, and formerly known as the Gilbert Islands) is looking for a new home for his 51,000 people, and no country is opening its doors. Kiribati won its first Commonwealth Gold medal at the Glasgow games in 2014.

Diseases carried by insects – like **the zika virus, malaria and dengue fever** – are on the rise in countries that previously only experienced them when travellers brought them home. Now mosquitos travel on ships and planes and survive in 'temperate' zones.

As the health and security of people worsens, they look for other places to live. The first specific environmental refugee was from Kiribati. He tried to immigrate to New Zealand, where his case and appeal failed. In 2009 the International Organisation for Migration studied forecasts for **environmental migration**. The projections ranged from 25 million to 1bn people, by 2050, with

200 million the most common estimated figure. In 2014 the International Displacement Monitoring Centre reported that 'natural hazards' displaced 22m people. This number was greater that year than migration caused by wars.

Insurance companies are moving out of high-risk markets. For example, if you live in an area prone to flooding, you may be hard-pressed to find cover for your home – and then only at high premiums.

The Sundarbans mangrove swamps protect Bangladesh from tropical cyclones (hurricanes). If the Indian Ocean rises by 45cm, 75% of the wetlands will lie underwater. The first Sundarban island was lost in 2010, and a 45cm rise could happen as early as 2050. When Bangladesh loses the swamps, it will **lose its protection** against storms and flooding. Undeveloped coastal regions around the UK offer the same benefits.

Not all areas experience sea level changes in the same way. If the land moves, rising waters may be offset or worsened. For example, Alaska's icy coat holds it down. **As the ice melts, Alaska rises**. The rising west coast of North America forces down the east coast (like a seesaw). As a result, sea levels on the Alaskan coast will rise less slowly than average; they may even drop while cities like Boston and New York will see greater rises. The same effect flooded some ancient Mediterranean ports and raised others, and it created Finland. In the UK, Scotland is on the up while England is sinking (our glaciers melted about 8,000 years ago).

Australia will see average temperatures rise up to 4.5°C by 2070. Parts of the country are already in drought and experience bushfires outside the 'normal' season for such events. Africa will continue to get drier. Asia and South America will experience **greater weather extremes**, as the frequency of the Southern Oscillation (El Niño and La Niña) increases in response to higher temperatures.

In 2012 the world's fisheries caught 66.6 million tonnes of food fish and 23.8 million tonnes of aquatic algae. Not only did they feed millions of people they added US$144.4 billion to the world economy. However, between sea temperature rises and ocean acidification, many ocean species are under **threat of extinction**, and with them our food supply.

That fracking debate

There is no doubt that fracking is one of the most contentious advances in oil technologies and has given rise to a furious public debate. This section touches on each claim and counterclaim. But first: some observations on the debate itself.

Each side uses arguments that distort information about fracking as a whole. Anti-frackers employ one unique technique: They assess the oil and gas production cycle and assign its flaws to just the fracking stage; for example, they object to the noise and lights that come with all drilling and extraction while objecting to 'fracking'. Pro-frackers have two unique techniques. They narrow the argument to purely fracking and not its direct consequences – so, for example, some claim that flowback water disposal is not a fracking issue. And they lump all opponents into an amorphous blob. If the arguments of two people, or groups, contradict, they cry 'inconsistency'.

However, the two sides have a lot in common. Each side uses science showing (not necessarily relevant) points to enhance their arguments. They take single points of data to make emotive assertions (typically, environment and health on one side, and economics on the other). They exaggerate the other side's position and then use data to show it is much worse or not as bad. Both resort to unpleasant tactics, such as personal and generalised abuse. In drawing your own conclusions about fracking, do not be swayed by the way the two sides argue. Think about what you want to know and seek reliable data.

Claim – fracking causes earthquakes

For:

- Shale gas operator Cuadrilla acknowledged that shale gas exploratory drilling caused two earthquakes off the north-west coast of England (Blackpool).
- In geothermal energy production, fracking has caused earthquakes in Switzerland and California (generally through pumping water).
- Disposal of fracking wastewater in Ohio causes earthquakes.

- Fracking has caused a 30-fold increase in the number of earthquakes in the US.

Against:

- Very few fracked wells have led to earthquakes.
- Proven quakes are small in magnitude – between three and four on the Richter scale.
- Governments have improved regulations.

Claim – fracking uses too much water

For:

- Fracking uses a lot of water, which comes from the mains, from rivers or water table extraction, thus depriving others and the environment of its use.
- Fracking has pushed some US towns into drought.

Against:

- The industry consumes a tiny proportion of water (0.3% of the water consumed in the United States). The volume is less than the water used to keep golf courses green. Similarly, you could only make 701 cotton T-shirts with the water employed in the average fracking operation.
- The industry has learnt how to use salt- and wastewater.
- Gas-fired electricity generation uses less water than coal and that offsets some of the water used in fracking.

Claim – flowback[191] water is hazardous

For:

  T Flowback water has caused pollution when holding ponds leak or are subject to flooding.

Against:

  T In the UK, Environment Agency-approved processing facilities will clean the water. This capability already exists, processing wastewater from the North Sea and mines.

Claim – fracking chemicals[192] are hazardous

For:

  T The anti-fracking lobby says that only transparency is good enough to understand the hazards.

Against:

  T Cuadrilla has only used three chemicals in the UK: polyacrylamide friction reducers, commonly used in cosmetics and facial creams; hydrochloric acid, frequently found in swimming pools and used in developing drinking-water wells; and biocide, used on rare occasions when the water needs to be purified.

  T The Environment Agency will not permit the use of chemicals that are hazardous to groundwater.

Claims – methane leaks negate the greenhouse gas emission benefits of displacing coal

For:

  T Studies have shown that gas infrastructure leaks are greater than disclosed by the gas industry or the US EPA.

  T Studies have shown that leaking well casings allow methane to escape.

---

191 After fracking, some water stays in the ground; the rest 'flows back' and is captured.

192 The exact makeup of fracking fluid is unknown, as the United States has no requirement for companies to reveal their chemical blends.

- Studies have shown greater levels of leaks from natural gas operations than disclosed by operators or the US EPA.

Against:

- Scientists have shown that fracking is highly unlikely to release methane into either water reserves or the atmosphere.

- The United States EPA states that the methane released in natural gas operations does not offset the benefits of burning natural gas instead of coal.

- Some studies about leaks from natural gas operations (not infrastructure) are partisan.

Claim – fracking contaminates groundwater

For:

- There are highly publicised claims, most famously through the movie *Gasland*, of flaming faucets and foul-tasting water.

Against:

- The US EPA released a (draft) report showing that fracking has not 'led to widespread, systemic impacts to drinking water resources' though it also 'identifies important vulnerabilities in drinking water resources'. Though the conclusions of the report have been disputed by the agency's own scientific advisors.

The Administrator (head) of the United States EPA, Gina McCarthy, told *Boston Globe* reporter David Abel in November 2013 that sound engineering practices make fracking safe. We should hope the industry universally adopts sound practices though perhaps some players will need regulation. While improved methods will reduce the environmental impacts of fracking, costs will increase.

Prevention is better than cure

People often respond to calls to reduce pollution with 'there's no point in me reducing, if everyone else carries on'. Well, you have seen that others are no longer burning fossil fuels. If you reduce your use, you will join a growing group.

Use less

If we want to use oil, we must stop wasting it. We have to squeeze from it all the value we can and dispose of the remains without littering our home. We have already thought about energy, so let's consider two other things we waste: food and materials.

Every year UK households throw away 7m tonnes of food and businesses chuck out 8m tonnes. Earlier we saw that the average human eats 387kg of food a year, so our annual food waste could feed 39m people. If we look globally and include waste in the supply chain, the numbers get even worse; we throw away one-third of the food we produce globally.[193] At the same time, around the world, one in seven people cannot get enough food. We, the rich countries, could save the starving billion without tightening our belts. Indeed, if we stopped throwing away food we could feed the hungry and decrease oil use.[194]

We know how to use fewer materials. We could shop less and be selective in what we buy, choosing items with longer lives and from sustainable sources. To do the latter, we need help from business, as the key to more sustainable products is design. Good design reduces the materials we use. It increases the life of products by making them more hardwearing or easier to repair. Thought in design makes products oil-cheaper to make and transport, and it lets us recover materials more easily. The companies that embrace sustainable design reduce costs, attract new customers and develop business models based on products and services, rather than focusing on just making more.

The more we instil sustainability into our lives, the less oil we will use and throw away. A good example of sustainability working (though not explicitly) comes from America. Shale gas supporters point to falling American carbon dioxide emissions in 2012. They would have us credit shale gas with the full 3.8% drop seen in 2011-2012. They make a similar claim for the 12% 2007-2012 carbon emissions decrease. However, a closer look at the data reveals other contributors:

---

193 This total does not include the 90% of caught marine life discarded by fishermen.

194 After a difficult legislative birth, French supermarkets are now legally obliged to prevent food waste or, if they cannot, donate or reprocess the food. If that is not possible then it must be recovered for animal feed and, if all else fails, it can be used as compost or in energy recovery.

⚡ Lower energy intensity. American GDP increased by 2.8% in 2012 and energy consumption fell by 2.4%. Using less energy delivered a reduction in carbon dioxide emissions of 282 million tonnes.

> *Energy efficiencies in vehicles accounted for 22% of the decline in emissions. Vehicle miles were stagnant.*

> *Due to mild weather, more than half the energy intensity decreases came from reduced demand for domestic heating oil and electricity.*

> *Reduced electricity demand and reduced electricity losses made up the remaining reduction in emissions.*

⚡ Lower carbon intensity. Using fuels that emit less carbon dioxide accounted for 75 million tonnes less carbon dioxide.

> *The drop in carbon intensity in 2012 is entirely because of the switch from coal to natural gas. Non-fossil fuel sources contributed no decrease overall, as hydroelectricity production fell more than new wind and solar energy grew, and nuclear was static.*[195]

Looking at these data graphically shows that warm weather contributed most to the 3.8% annual reduction in US carbon dioxide emissions, and both electrical energy efficiency and vehicle efficiency gave the dash to gas a good run for its money.

In the longer term (2007-2012), US emissions have dropped by 12%. As the chart below shows, the lion's share of the decline came from a reduction in energy intensity.

⚡ Energy intensity reductions during this period had four main drivers:

> *Weather (warmer winters, cooler summers) – uncontrollable;*

> *Reduction in industrial and manufacturing outputs – not good for the economy;*

> *Reduced vehicle miles;*

> *Improved vehicle efficiencies.*

> *The carbon intensity of electricity production from 2007 to 2012 dropped by 13%:*

> *8.2% (or 198 million tonnes less carbon dioxide) came from the shift from coal to natural gas.*

---

195 Hydroelectric electricity production depends on rainfall and melting snowpacks, and varies from year to year.

> ➤ *3.8% (or 116 million tonnes less carbon dioxide) came from a 9% increase in non-carbon generation (renewable and nuclear).*

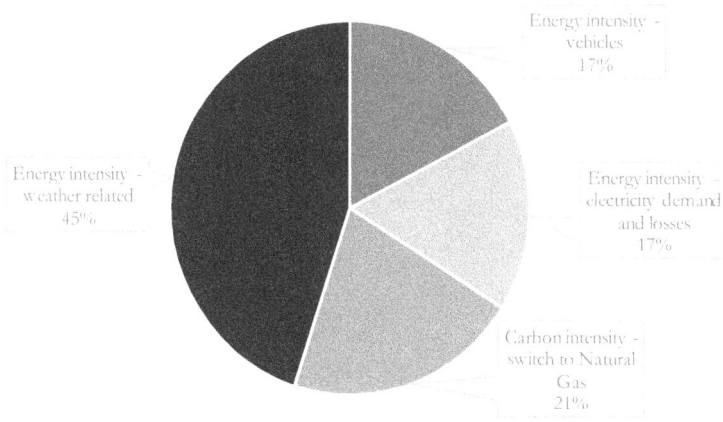

Figure 26: The drivers of the drop in US emissions – 2011-2012

You might wonder if there was the same fanfare over the 2013 figures. There was not. The EIA analysis for a single year is statistically complicated; it reports the emissions over the trend of the previous decade. On that basis, emissions in 2013 grew more than the trend in all areas but one. Energy intensity – the amount of energy used for each dollar of GDP – rose. The main reason energy intensity increased was the freezing 2012-2013 winter. The economy also grew, and emissions increased with it – and rising natural gas prices saw the electricity industry turn to coal. The factor that did not increase emissions was population.[196]

---

196 Because the population grew more slowly than in the previous decade.

Drivers of US CO_2 emissions reduction 2007-2012

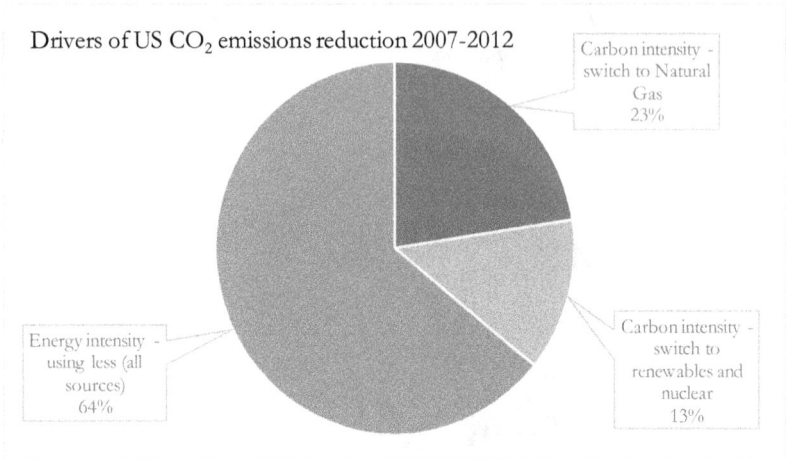

Carbon intensity - switch to Natural Gas 23%

Carbon intensity - switch to renewables and nuclear 13%

Energy intensity - using less (all sources) 64%

Figure 27: The drivers of the drop in US emissions – 2007-2012

So has shale gas reduced carbon emissions? Yes. Is it accountable for the 2012 reduction? No. Can we depend on shale gas to sustain a reduction in emissions? Not in a free market, where changes in energy source meet each price fluctuations.

Do these data show us how to reduce emissions? Yes. While we cannot control the weather, we can insulate our homes and offices against cold and hot weather, use better heating and cooling controls and drive more fuel efficient vehicles. Improving energy efficiency permanently would emulate and beat those US successes.

Capture

If we are to reuse, recycle or store the products of oil, we must recover them. When we considered plastic recycling, we saw that the onus of recovery has moved away from the consumer. Instead, companies use technology to separate plastics.

Taking that one step further, we find businesses that specialise in gathering valuable materials from landfill, road sweepings and mine tailings. They extract the gold from car exhausts, from road debris and, perhaps, from waste water. While some find metal among muck, others generate value from electronic waste. These new industries keep valuable resources in use instead of burying them. With them, we can reduce mining and quarrying.

Again, design comes into play. In "Plastic dolls" we thought about active disassembly, which only works for electronics designed with materials-recovery in mind. Thank goodness for the schools and universities teaching the designers of tomorrow.

As we burn fossil fuels in power stations, we release gases in a fixed location, making them easy to capture. Carbon Capture and Storage schemes are popping up all over the world. (In America the same concept is called Carbon Capture and Sequestration. Handily, both are known by the initials CCS.) As you might expect, there are several schemes in America, where different methods are being tested and assessed. Unfortunately, a change in UK government policy has pushed out the biggest investor in our most advanced scheme.

Whether we store carbon, often pumping it into spent hydrocarbon reservoirs, or use it as raw materials, CCS has two disadvantages. The first is that it seems to give us the freedom to continue burning fossil fuels. From an environmental (and oil-price) perspective, that wouldn't be good, because the effects of using fossil fuels do not limit themselves to climate change. Second, we do not yet know how to bolt CCS onto existing power plants. At best, very expensive refurbishment is required; at worst, brand new power stations are built – with all the oil costs of construction and demolition (if the old plants don't just carry on churning out CO_2).

At least capturing carbon from fixed locations such as power plants is relatively straightforward. Capturing the gases resulting from transport is more difficult. Never fear: scientists are working on methods of taking carbon from the air. They call this form of capture Indirect or Ambient CCS. Ambient technologies can be a scaled-up version of chemical scrubbers, use catalysts, or be living. The irony of the last type is that we might destroy desert ecosystems to catch the carbon dioxide released (in part) by deforestation. Ambient carbon capture brings another problem: How would we know when to stop? How tempting to try to remove decades-old carbon. Do we understand the environment well enough to attempt this kind of geoengineering?

The reusing, repairing society

Another way of using less is to recycle (to restore a material to its virgin or near-virgin condition). It is a useful way of making the most of goods and materials but uses a lot of energy. 'Upcycling' (converting used material into something with more value) is largely a niche activity, dependent on human labour and commensurate pricing. What's more, both condone our disposable habits. Repair must come before we replace our goods hoping someone else will recover the value from our scrap. Our parents and our parents' parents mended their belongings. They kept pieces of string, old screws and buttons just in case, and we have let oil rob us of that practice.

Oil lends convenience. When an oil-based product breaks, you replace it. Indeed, why wait for breakages? If your stuff bores or embarrasses you, or you can't be bothered to hold onto it until you next need it, a replacement is just around the corner. With oil, we no longer have to look after our stuff or repair it. Why should we? Our lives are busy, and things are cheap.

We no longer 'make do'. Our personal, business and social lives do not feature the time, skills or inclinations. To turn things around, we need to learn how our stuff works. Sewing on a button, fixing a broken lamp or replacing a phone screen could be the difference between having and having not.

However, individuals have a limited role in developing a 'repairing society'. We need to make repairs ourselves, or be ready to pay for repair services, but business and government must act. Manufacturers often do not sell spare parts, and we need access to restricted, specialist knowledge and equipment to take apart safely and reassemble products. To advance from disposal to repair, we need items *designed* for repair, and for repairs to be worthwhile. Where short-sightedness currently prevails, governments can introduce standards for reparability, and penalties for companies whose products cannot be fixed.

We need to repair outside the home, too. In the office and factory, we replace items we could otherwise clean or repair because our economy makes it cheaper to throw them away. We act wastefully because the expense of repair seems greater than the cost of the waste. But as the price of wasting materials rises, we will wish we had invested when we could. If you expect some examples of repair and reuse about to follow, you will not be disappointed.

Many specialist businesses now repair equipment and recover materials, and more join their ranks weekly. The following gives a tiny cross-section:

One company uses spiral welding to **replace the material on worn shafts**.

Another **recycles hybrid car batteries** to extract lithium and other valuable materials.

An innovative student from the University of Brighton **makes new plastic** from discarded chewing gum.

A UK manufacturer accepts electronic waste from its customers and uses it as a **source of parts**.

Do you remember the company in "Oiling the wheels of industry" that **presses new parts** from the waste sheet metal of other manufacturers?

Some councils **refurbish their dustbin lorries**, instead of scrapping them and buying new.

Ecover, which does not use petrochemicals in its washing products, has started to use **plastics from the sea** to make its plastic bottles.

A Swedish company will sell you a pair of jeans, repair them free of charge and, when they are beyond repair, you can **trade them in**. They use old jeans to provide repair materials.

In South London you will soon be able to visit the **Library of Things**. And not just visit, you will be able to borrow *things* for up to a week. From suitcases to power drills, the Library of Things lends out stuff you would otherwise buy, use and then stick in a cupboard.

We have seen a funky method for disassembling mobile phones. How about recycling the circuit board too? At the moment, we can recover about 2% of the materials on a circuit board. If we change the disassembly method by adding a giant dishwasher and **glue that dissolves in hot water**, we can up the rate to 90%. Three UK firms got together to develop a glue that holds electronic components together during use, so this is now possible.

Even better, why build mobiles that will become obsolete? Two Northern-European start-ups (one in the Netherlands and one in Finland) are launching phones that **anyone can repair or update**.

The owners of the **Empire State Building refurbished** the historic building from 2010 to 2012. The project has reduced energy costs by 38% ($4.4 million) a year. And inspired other building owners to follow suit.

In August 2013, the **White House was treated to new solar** (photovoltaic) panels and energy efficiency measures, including new controls and variable-speed fans. The improvements will save 19,700kWh a year and will pay for themselves in eight years.

In new buildings, and in refurbishment, we could use **concrete made by microbes**, from sand and the nutrients in urine. Or mimic the way marine creature make limestone.

The owner of a ski shop in Whistler, Canada and a customer got talking about ski boot liners and how they went to landfill after use. Some heavy thinking later, the customer set up a non-profit organisation that **fits used liners to rubber boots** and ships them to Mongolia. The liners no longer clog up landfill and feet in Mongolia are much warmer.

An academic, business and governmental consortium in Wisconsin are working to **reduce the environmental impacts of dairy farming** and make individual farms less dependent on external (oil-based) products and services.

A professor and several enterprising students at the Colorado School of Mines have shown that ash from food-waste incineration contains minerals essential to the production of materials such as glass. They have made the latter **out of banana skins**.

A French yogurt maker aims to produce **zero net carbon emissions** by 2050 and will tackle energy efficiency, water- and waste-management, and farming practices along the way.

There is a coffee shop on virtually every street corner; imagine the colossol piles of coffee grounds that accompany each one. Most of us don't give a second thought to these virtual, smelly heaps, but for one British designer they are a source of materials. He has set up two organisations to use the waste. The first turns it into **fertiliser; the other makes furniture.**

A US ice-cream company collects the little plastic spoons they dole out with each tub and turns them into the **filaments for 3D printing**.

If your best-loved shirt is showing signs of age, you can have it converted into **boxer shorts** in a Dutch social enterprise, which helps women make their way into employment.

"Nuts, whole hazelnuts…" Nutella is the largest consumer of hazelnuts in the world and, as they cannot use whole hazelnuts, they have also been the biggest disposer of shells in the world. Well, no more. It is impossible to tell apart card made from hazelnut shell fibres and traditional cardboard. Other nut shells are also handy as they contain cellulose, which can provide feedstock for bioplastics.

If you are short of nuts, but grow tomatoes you can use **dead plants** to enrich old paper and make strong cardboard boxes (to hold tomatoes).

A company in Wales turns **old carpets** into flock and the fluff is used as a surface to protect horses in training.

Sailors tired of the mess of plastic in ports, harbours and on beaches have designed the deceptively simple **seabin**. It sucks up water and separates the debris, leaving water in marinas clean and safe for wildlife.

Do you remember the call of the rag-and-bone man? Commercial, social and government enterprises are emulating this age-old role with so-called **'pop-up' recycling**. Just like their forebears, modern-day rag-and-phone collectors encourage householders to give up their reusable and recycling materials – though today, the reward is a cleared-out loft rather than cash.

Of course, if you can repair, then pop-up recycling becomes less necessary. **'Makerspaces', 'hackspaces'** and a host of other great names are places where people can go to repair stuff. Machines, advice and likeminded souls are there to help anyone make the most of their belongings.

You can see that the actions required to look after our planet also protect our economy. However, many businesses are only profitable if we replace old, broken items. If we wish to make repairs a way of life, we have to change business models. We have to design for repair. Products must be modular – easy to disassemble and reassemble. We need good repair stations, spare parts and test equipment. Companies must collect data about repairs and faults, to improve future designs. And we have to be willing to pay.

Chapter fifteen

To oil, or not to oil?

THIS CHAPTER IS ABOUT GETTING ON WITH IT. We discuss who should lead the move from oil and how we can be helped to do so, and we take inspiration from some historic changes.

If we accept that oil prices are undependable, we need to take action. Sure, some people need to do some grand strategizing, but, for the most part, you and I can simply get on with removing oil from our lives – we have already seen how. Yet we leap to defend our use of oil. We complain that others are not changing, and ask why we should. We employ double standards. We spend more than we can afford to keep up with the Joneses and refuse to pay a little extra for something that will protect our world.

Would one or two little changes be that hard to make? They would quickly add up, and each brings the benefit of making the next even easier. The big question is: Who should lead this change? For the answer, look in the mirror.

We the People

We are often so busy thinking about the things governments and businesses should do that we underestimate the difference we can make. You have all the power you need. Senator Albert Jeremiah Beveridge of

Indiana coined the term 'grassroots', saying of the Progressive Party in 1912:

> *"This party has come from the grass roots. It has grown from the soil of people's hard necessities."*

Change has grown from grass roots in extraordinary circumstances. Grassroots organisations gained women's suffrage. Workers built the trade unions. Farmers and consumers grew the organic food industry. Whether you support these ideas, you cannot deny that grassroots efforts have made changes. You can join with like-minded others to demand new products and services that reduce dependency on oil. You can tell the people you elect that they must act. You can make changes to buffer you and your family from the effects of undependable oil prices.

Okay, we can't do it alone. You know business and government are watching you. They are not engaging in voyeurism; they want you to be happy with their actions (so that you will buy more, or re-elect them). If you ask for it, businesses will turn their creativity to making the alternatives to oil. If you ask for it, the government will support the development of alternative technologies and business models. If you ask for it, we will stop paying the price of oil. Getting started will quickly signal your intent. With luck, the examples in this book have sparked your curiosity. You may be wondering whether you can make changes now. You can. Here's how:

> Learn more about oil and demand clear, **open debate**. Use social media to ask questions and get good-quality answers, and always ask for (good quality) data to back up an opinion. Do not let anyone fob you off with complicated answers full of jargon or rhetoric; insist on evidence and ask the difficult (even if seemingly stupid) questions. Don't let people distract you with *reductio ad absurdum*,[197] over-simplification, [198] a newspaper's inaccurate

---

197 Taking an argument beyond its logical conclusion – eg 'If smog kills, then why isn't everyone in China dead?'

198 Just taking one part of a complex situation to disprove a theory – eg 'Look, ice grew quickly in the Arctic even though this year was the warmest on record – therefore, temperatures do not impact ice growth!'

report of a science story, [199] conflation[200] or, indeed, accounts of old claims that have since been proved wrong eg 'global cooling'.[201][202]

Discuss **what you have learnt** down the pub, at the school gates, in the coffee shop, at work, at the gym, with your investment adviser, at church, mosque, temple, synagogue or chapel.

Take inspiration from the millions of people who use renewable energy or who have employed efficiency to reduce their energy, food and shopping bills.

Do not sit back and let others decide for you. A few small steps taken by each of us will start the removal of oil from our economy. Ask your supplier for renewable energy. Invest in micro-generation in your home, business or community.[203] Reduce the energy you use. Improve insulation. Travel in different ways. Cut down on the plastic you use and throw away. Write to your paper, speak with your MP, and petition your Energy Company and supermarket.

---

199 Unfortunately, there are times when journalists have misinterpreted a scientific finding and published sensational news (eg science says winters are getting warmer; newspaper reports the 'end of winter').

200 Combining two pieces of information that seem related in a manner that makes two plus two come to the answer I want instead of four.

201 Why? If you read the appendix about climate change then you know there is a consensus among climate scientists; to get to this point there has been lots of data-gathering, analysis and hypothesising. This is part of the scientific process. Naturally, some hypotheses are proved wrong with more data or just time. The fact that scientists have been wrong in the past shows this subject has been carefully considered.

202 Throughout my research I have checked, double-checked and, sometimes, triple-checked the spin people put on data. Thankfully, there are some excellent resources that help with this activity. *Without hot air* takes the hyperbole out of the argument for renewables and *Skeptical Science* presents cogent debate about climate change. *Climate Feedback* links feedback from scientists to newspaper articles. You will find links to all of these in the Sources.

203 If you go ahead and buy renewable energy capabilities or invest in peer-to-peer lending, please be careful. There are bad people out there willing to rip you off instead of providing a good product or service. Research the companies you consider, ask for reviews or to see their work, and always check business credentials.

Consider investing in renewable projects and divesting[204] from fossil fuels.

Repeat these actions until you are happy you and yours will be less vulnerable to the vagaries of oil prices.

It is tempting to wait; after all, there are new technologies announced every day, prices fall year-on-year[205], and no one wants to get stuck with the Betamax of energy generation. Today, right now, you can save money. Today, right now, you could be smug when the electricity goes off, or gas prices rise. Today, right now, you should take advantage of the support offered by the government, and signal to business and government that you want more.

Show me the money

Without a doubt, humankind's ingenuity and industry got us where we are today. Given the right attention, we can work our way through and beyond our addiction to oil. But this does require money up front. Few companies have cash sitting in the bank, so most need to raise funds to transform their business. The most common ways of raising funds are through shareholders or loans. Shareholders receive dividends as a reward for the risk they take, and loans accrue interest. Therefore, each investment must improve the business' finances, and this has to be proved in advance of any money changing hands.

Most people understand the idea of payback. To assess the value of an investment, we work out how quickly savings will outweigh costs. Let's take a home-based wind turbine that reduces an electricity bill by £70 a year and costs £500 to install. The payback would be 500 divided by 70, or the best part of seven years and two months. However, the simple payback calculation cannot compare complicated investments, so businesses use a calculation called Internal Rate of Return (IRR). It works

---

204 First and foremost, this is not financial advice. Second, what is 'divesting'? The divestment movement is encouraging large investors – eg universities, councils, pension funds – to sell their shares in fossil fuels (and other unsustainable industries). Some notable institutions have done so, including the Church of England who pulled out of an oil company planning to explore for oil in the Virunga National Park, DR Congo.

205 From 1970 to 2013, photovoltaic prices dropped from $76/kW to 74 *cents*/kW (US$). Forget the units –admire the scale. And in 2016 two groups of researchers (from the NOAA and Oxford University) independently predicted even greater advances.

out the financial benefit of an investment and presents an equivalent interest rate. Investors compare the interest rates of investments to each other and the cost of the money. (Though the calculation you use does not matter if you cannot see the numbers in the dark.)

That last point demonstrates the problem with purely financial assessments. Although they calculate the costs of staying the same, they rarely allow for the kinds of change in circumstances we can expect. So, strong business leaders consider all alternatives and add them into the mix. They keep an eye on the future to make sure the company they manage survives. Shrewd leaders take their businesses through redefinitions. For example, some maritime companies are driving ahead with changes that are not yet required by international law. Therefore, here and there you will find companies driving low-carbon agendas.

Here's a question for you – ten points if you correctly identify the company described by its Chief Supply Chain Officer below (answer in the Sources):

> *"Across our supply chain we are increasingly turning to energy provided by wind, solar and biomass, converting heat from our manufacturing processes into power for our factories. We are on track to reach our target of 40% renewable energy by 2020. [And carbon positive in our operations by 2030]"*

Making this kind of change needs belief. Unfortunately only half the world's CEOs believe climate change is a threat (and even fewer appear to worry about the other subjects we have discussed). We must raise our voices to inspire further strong leadership, creativity and courage.

Not for profit

There are many ventures, from the social enterprises we met in "It's the economy, stupid" to charitable organisations, who will fund technology development and implementation. For the most part, they will not contact you – you will need to hunt them out, but if you do, you will find guidance and support on a scale you probably hadn't imagined possible.

Democracy gives us the right to fail to prepare for the future

"Democracy gives every man the right to be his own oppressor."

– James Russell Lowell

Wishing someone a role in government is like cursing them; fortunately, some people put themselves forward for this experience. Imagine walking a tightrope between imposing a nanny state and letting it all go to h*ll in a handcart; would you like to do it? No – it sounds too much like hard work, so let's agree: no politician-bashing.

If we are not going to bash them, you might ask whether they support our move from oil. We will see in a second that the UK has legally-binding obligations to use less oil. Further, we wish to have energy with less in the way of international politics (Americans call it 'energy independence'). And many politicians realise that burning oil compromises our health and economy; elected officials and civil servants are already preparing for or managing the effects of pollution and climate change. Even better some of our leaders and would-be leaders have noticed that new, related, industries, from solar to waste management, are driving growth and job creation.

As a result of all this, the government looks to be taking action to reduce our dependency on fossil fuels. But saying 'We will reduce carbon dioxide emissions' is not enough. A government intent on moving away from oil leads by example. It pushes energy efficiency in public buildings. It removes petrol and diesel vehicles from its fleets. It helps everyone else stop burning fossil fuels. Setting and delivering these kinds of policies has two benefits: a reduction in our dependence on fossil fuels, and stable markets for innovators and businesses. The latter makes funding for product development and business growth more likely, and cheaper.

A government resolved to reduce the risks of undependable oil prices sets policies that promote change. They don't have to be controversial or difficult policies; the IEA recommends a simple set of five:

1. *Increasing energy efficiency in the industry, buildings and transport sectors*

2. *Reducing the use of the least-efficient coal-fired power plants and banning their construction*

3. *Increasing investment in renewable energy technologies in the power sector from $270 billion in 2014 to $400 billion in 2030*

4. *Gradual phasing out of fossil-fuel subsidies to end-users by 2030*

5. *Reducing methane emissions in oil and gas production*

So what we need are policies (and supporting actions) to encourage take-up, support nascent industries, and drive research and development. We are about to look at the UK government's actions and achievements in each of these areas, and will consider how well it is acting as a role model. We will end our dabble in government by looking at how international co-operation helps us adopt a less oily way of life, and the actions other countries are taking.

Local vs national government

In the UK, the two forms of government are very different beasts. They differ in their accountabilities and ways of working. However, they are both necessary in the transformation we need to make, and their roles are similar. For simplicity, this chapter only distinguishes between the two types of government in particular examples.

How are 'they' doing?

An independent assessment of our performance against our European commitment, [206] in June 2015, had two sobering messages. First, our targets are among the lowest in Europe, and second, while we have hit the government's interim targets, we are unlikely to achieve the goal of 15% by 2020. By February 2016 we were the third slowest EU nation to decarbonise. What's more, in October 2015, Amber Rudd (Energy Secretary) wrote to her Cabinet colleagues stating that the absence of a credible plan to meet the target carries the risk of successful judicial review. This statement can be translated into plainer English: Unless the government puts a credible plan in place, we will be fined by Europe.

More recently two other bodies have weighed in to criticise successive government's strategies for energy. The Institution of Mechanical

---

206 The EU targets are 20% of energy from renewables by 2020 and 27% by 2030. The UK adopted a 15% target for 2020, consisting of three sub-targets: Transport 10%, heat 12%, and electricity 30%.

Engineers has said there is a move away from top priorities, and the Siemens' director of energy strategy and government affairs told the Energy and Climate Change Committee "That's no way to run the strategic energy policy of the country". So what does the government need to do?

ACT AS A ROLE MODEL

Government must practise what it preaches and, to some extent, the UK government does:

> The 2010-2015 coalition government made a commitment to **reduce the carbon footprint of central government** offices by 10% within one year in May 2010. The government exceeded its efforts and achieved a 13.8% drop, which reduced the government's energy bills by £13 million. Out of interest, the three runaway winners were the Ministry of Justice, the Department of Work and Pensions, and Her Majesty's Revenue and Customs.

> To build on the successes of the one-year target, the government made several **Greening Government Commitments**. These took in a broad range of objectives, from reducing greenhouse gas emissions to reporting progress in annual departmental accounts. Its achievements are impressive and could, if more widely known, inspire other large organisations to do the same.

> In 2012 the Local Government Association set up the **Climate Local initiative** with support from the Environment Agency. It helps councils reduce emissions and become more resilient to changes in our environment (eg by implementing flood protection). 96 local authorities signed up by April 2014. These councils (and unitaries) have reduced carbon emissions from their buildings, helped local people take up energy efficiency measures, supported local businesses in the low-carbon economy and set up energy generation companies.

ENCOURAGE THE RIGHT ACTIONS

There is a carrot-and-stick approach to getting us to turn from oil. The carrot offers incentives and support. The stick uses legislation and financial measures.

At a personal level, our carrots come in the form of activities such as the Green Deal and Affordable Warmth Obligation. Both of these assess our needs and help us get the right work done to reduce our oil consumption. And, as we have already seen, some local authorities are setting up generation companies to help those in energy poverty keep warm. But really, the support comes from the government's massive annual budget (c.£735bn in 2014-2015). This money is spent to encourage take-up and support industry so we will cover it in the next section.

The 'stick' of legislation comes in many forms. Our grandest is the Climate Change Act 2008, which mandates an 80% cut in six greenhouse gases against 1990 levels by 2050 (we were the first country to introduce legally-binding targets).[207] Because it is a law, the government has given itself a pretty big stick. The legal consequences of failure are not clear.

There is more legislation and regulation to coax government and businesses into action (if they are smart, they also take the help offered at the same time). For example:

The operators of frequently visited public buildings which have a floor area greater than 250m² must put up a **Display Energy Certificate**. In compiling the information for these certificates, officials can quickly see where they have opportunities to save energy and money. Unfortunately, there is no legal standard for public buildings, which means they can continue to be inefficient, as long as they show they are.

Public bodies that consume over 6,000MWh of electricity a year and participate in the half-hourly electricity market[208] have to report their emissions and buy allowances. If they reduce their consumption they can sell unused allowance (which can make money for them); better still, their charges for the **CRC Energy Efficiency Scheme** (for that is what it's called) are reduced. The

---

207 According to an EU roadmap, this target is more ambitious than the EU target mentioned earlier.

208 You may well ask. We have no idea how easy we have it, buying energy at home. We agree a tariff, use the stuff, get billed. If you're a major user of electricity, then you contract with a supplier to use a certain amount of electricity every half hour of the day. If you don't, there are financial shenanigans (called a settlement process) to make sure that no one is inconvenienced (say, through power cuts) or allowed to cause disruption (for example, by using more electricity than agreed). Businesses that generate electricity also operate in this market and engage in the settlement process.

same applies to private sector organisations who meet the same criteria.

The CRC Energy Efficiency Scheme is effectively a form of carbon tax; in the UK we also have two others:

> The **Climate Change Levy (CCL)**. All businesses pay the CCL unless HMRC gives tax relief. For example, energy-intensive industries can pay reduced rates if they meet challenging emissions-reduction or energy-efficiency targets agreed with the Environment Agency. In 2012-2013 the CCL raised £635 million. It excluded fossil fuels used to generate electricity and oil, as the latter attracts fuel duty. The government redistributed the receipts from the tax through a decrease in National Insurance (which has since increased again).[209]

> **Carbon Price Support (CPS)**. This tax forms part of the Carbon Floor Price and applies to the fossil fuels used to generate electricity. Introduced in April 2013, the combined tax take for CPS and CCL was £1,068 million in the 2013-2014 tax year. HMRC suggests that the bulk of the increase comes from the CPS portion.

Together these carbon taxes are called the UK carbon price; they encourage investment in low-carbon generation and accelerate the removal of fossil fuels from our energy mix. Our carbon price operates alongside the European Union carbon-trading scheme.

Taxes generate income for governments but are unpopular. To make carbon taxes more acceptable, some governments use the income to offset regressive taxes. They call these schemes 'tax neutral'. By reducing National Insurance, the government intended the CCL to be tax neutral – or at least not too bad.

So much for financial measures – what about laws? Governments can make laws that are beneficial and to send a message. For example, legislation to stop us throwing away valuable materials would benefit the recycling industry and ensure we keep rare materials circulating in the economy. However, the UK Government prefers voluntary measures so we cannot expect this kind of action.

---

209 You should note that since July 2015 producers of renewable energy also have to pay the Climate Change Levy (they'd call this a 'head scratcher' across the pond).

CHOOSE THE INDUSTRIES TO SUPPORT

Subsidies are an interesting subject – in a data-driven world you would
expect to click a few links and get clear, comparable numbers that let us
understand where the government is putting our money. Alas, we have
some way to go before we reach that particular nirvana. So this section
will compare the fossil fuel subsidy figures calculated by the OECD to
the big three support mechanisms for renewables and the annual reports
of the UK governmental departments most closely related to each
chapter in Part One.

Even though the government doesn't collate data about its spending
on supporting renewable energy and un-oily materials, it has recently
revealed the cost of the three biggest subsidies for renewable energy.
They are:

The **Renewables Obligation** (RO) – the government puts an
obligation on all electricity suppliers to provide a certain amount
of electricity generated by renewables. At the same time, the
government gives all accredited generators Renewables Obligation
Certificates (ROCs), to show they have produced electricity from
renewable sources. To show they have supplied enough renewable
energy, the suppliers buy the ROCs from the generators and then
show them to Ofgem. If the amount of renewable electricity
supplied is less than the target, the suppliers have to buy additional
ROCs. The money paid for the ROCs supports the generators of
renewable energy and comes from our energy bills. This measure
cost the energy companies £3.5 billion in 2014-2015 and will rise
to £9.1 billion in 2020-2021. The RO is being replaced by
Contracts for Difference and will close to new generators from
March 2017. Solar was removed from the RO in December 2015.

Feed-in-Tariffs are also paid out of our energy bills. They have
two components. Each small-scale generator can claim a payment
for all they energy they produce – even if they use it themselves.
The second part of the scheme pays the generator for the
electricity they export to the grid. FiTs cost us £740 million in
2014-2015; the cost will rise to £1.6 billion by 2020-2021.

Contracts for Difference (CFD) – this is the one you may have
heard about with relation to the Hinkley Point nuclear power
station. If you want a CFD, you have to demonstrate you will meet

fixed criteria and may have to compete for funds.[210] If you are successful, the government will pay any difference between the cost of generating the electricity and the market price. This scheme directly supports low-carbon generators. In 2014-2015 it cost £5 million and the cost will increase to £2.7 billion in 2020-2021. General taxation pays for the CFD.

Now we have to ask: Are 'they' (that is, individual government departments) using any more of the budget to encourage investments in low oil technologies and practices, and if so, how? [211]

Here there and everywhere – The Department for Transport (DfT) is working towards a number of goals, such as populating our roads with ultra-low emissions vehicles (cars, vans and HGVs); putting in place a charging infrastructure; making roads both more resilient to climate change and inclined to produce fewer emissions; and targeting funding to support research and development. They intend these actions to reduce emissions, improve air quality, create and safeguard UK jobs and encourage inward investment.

Oiling the wheels of industry – The Department for Business, Innovation and Skills (BIS) puts in place funding for research in universities; supports innovation centres; provides State aid for energy-intensive industries; and helps companies develop and implement strategies for growth in manufacturing and services, including the low-carbon sector. It also established the Green Investment Bank.[212]

Plastic dolls – The Department for Environment, Food and Rural Affairs (Defra) provides grants for, and audits, waste

---

210 The government wants the scheme to find its feet before starting competitions; when the competitions start, the price paid for the electricity will reduce.

211 Of course, there are more departments than the ones examined here, and each one does much more than this – we are only asking how they support us as we move on from oil.

212 What is the Green Investment Bank? It a government-owned bank whose sole focus is financing projects that contribute to the UK's environmental targets. It supports projects of many shapes and sizes. The government plans to privatise the bank and strip its obligation to invest in clean technology.

recycling work. Local authorities run recycling. Some continually seek improvement and support innovation, though others do not.

Taking oil my energy – The Department of Energy and Climate Change (DECC) runs the Renewable Heat Incentive (a government-funded subsidy), Renewables Obligation, Contracts for Difference and Feed-in Tariffs. It also introduced the Infrastructure Act to support shale gas and deep geothermal projects, invests in the energy infrastructure, drives measures to ensure energy security and a low-carbon future, runs a governmental carbon-offsetting scheme, and decommissions nuclear power stations.[213]

Your goose is cooked – The Department for Environment, Food and Rural Affairs (Defra) consults and advises on the effects of the release of genetically modified organisms (GMO) into the environment.

Read oil about it – Defra also ensures the provision of broadband to rural communities and businesses, while the Department for Culture, Media and Sport (DCMS) is rolling out super-fast broadband.

Homeward bound – The Department of Health is reducing both energy costs and carbon emissions; it is improving video conferencing to reduce travel, recycles 78% of its total waste, and offsets travel carbon emissions. Meanwhile, Defra has reduced barriers to entry in the water market to stimulate competition and encourage suppliers to develop cheaper and more sustainable practices.

Where has all the oil gone? – Defra responded to the EC's strategy for the circular economy, saying we prefer voluntary measures in the UK. The same department manages national-level mitigations to climate change, eg flood control. As we saw earlier, it is also cleaning up pollution. We have already covered the actions of the DECC in this regard.

---

213 Clearly, decommissioning nuclear power stations isn't helping us decarbonise; but, as it costs more than half the core department's expenditure, it is worth a mention.

So now we know that the government gives some support to non-carbon technologies. Are they hedging their bets and also supporting fossil fuels? Well, that's a moot question, as we saw vast sums of money subsidise fossil fuels in "It's the economy, stupid". The chart below shows the way subsidies have been paid through both tax expenditure and budgetary transfer.[214] To be fair, there are a lot of inherited liabilities, especially from coal, but the OECD also expect us to give financial support in the future for Ring-Fence Expenditure Supplement, Field Allowance, Pad Allowance for Shale Gas, Brownfield Allowance, Mineral Extraction Allowance and Abandonment Costs. These are mainly costs associated with supporting oil and gas.

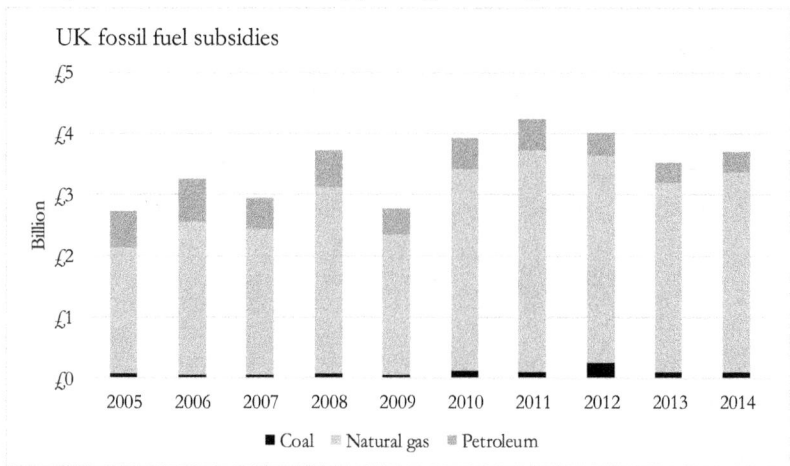

Figure 28: UK fossil fuel subsidies

DRIVE RESEARCH AND DEVELOPMENT

The media reports many amusing stories about pointless scientific research, and from this you might assume universities pursue only strange topics. Nothing lies further from the truth. Few studies ever make the national press, and the State rarely funds those that do. For example, Sky Travel paid for research to find the most miserable day of the year. State-funded universities have to show their research is sound to get government funding. One measure of success is the number of

214 In essence, the first is giving tax breaks and the second is direct expenditure. Successive governments have moved from the latter to the former.

peer-reviewed papers they publish. Another is the quantity (and quality) of references made to their studies in other peer-reviewed papers.

Scientific studies produce data which ensures debate reaches a sound conclusion. We use it to test theories, assess risks and make decisions. This book depends on data. Some sources of data are good. The World Bank's data site is excellent, as is the OECD's statistics site, both providing raw data measured in the same way across long periods and for most countries.[215] The United States Department of Energy's Energy Information Agency (luckily called the EIA for short) also hosts data on the Internet. The International Energy Agency (IEA) charges for most of its data. The UK and United Nations supply 'information'. The difference is: 'information' is data someone has interpreted. In the UK some sets of information change frequently. Sometimes the countries in the Union report using different methods. The landfill data in "Where has all the oil gone?" shows this best. Data is only available by nation, [216] but the measures are not the same across nations and data are infrequently collected. Without good data, governments cannot show the facts behind their decisions. We face difficult decisions, so a factual, data-driven, common understanding is vital. Fortunately, the UK government is improving data provision.

ARE WE ON THE PATH TO DECARBONISATION?

At a policy level, we learned in the autumn of 2015 that our government plans to phase out coal-fired generation. This move is dependent on our ability to shift to gas.[217] While support for offshore wind and nuclear continues, this policy change with the level of activity we have already considered would seem to suggest our government has not got behind removing oil from our lives.

---

215 But the World Bank calls theirs the World Databank, so they win on name.

216 Hence the sudden change from 'the UK' to 'England and Wales' in that section.

217 Another clear sign that the government intends to turn to gas is the decision by the Secretary of State for Communities to decide the outcome of Cuadrilla's appeal against Lancashire County Council's decision to decline planning permission for shale gas wells. Normally this would be decided by an inspector, but the Cuadrilla proposal has a significance greater than local considerations.

ONE MORE THING

Let's not underestimate the challenge faced by elected officials and civil servants. For example, if you were in charge, would you have banned the incandescent (traditional) lightbulb? Environmentalists favour the ban; they point to savings in energy and a slowdown in the increase of atmospheric carbon dioxide. Numerous groups and individuals fought the ban: some do not like the light quality of compact fluorescent lightbulbs (CFLs); others worry about the health effects of CFLs on people from epileptics to migraine sufferers and those with skin disorders. Some CFLs take time to light up and come to full power, compromising the safety of the infirm and visually impaired. Until safe disposal became available, CFLs were sent to landfill, from where harmful chemicals seep into the atmosphere, soil and groundwater.

More recently, opponents have complained about the timing of the ban, as it forced people to buy mercury-based CFLs. They say it should have been delayed a couple of years so that we could buy LED lighting. Another perspective is that companies would not have developed low-cost LED lighting without the guarantee of income provided by the ban.

And that's just lightbulbs. As yourself what you would do if you were in charge of any of these areas of government:

- Energy – invest in fossil fuels, nuclear, renewables or a mix?
- Food security – legalise drugs and take resources from food?
- Transportation – build a North-South high-speed rail link, upgrade existing railways or add more lanes to the motorways?
- Health – spend money to reduce the cost of energy, or only on health professionals?

We choose how much money we give to the government and how they spend it – at least, that is the idea of democracy. But the politicians we elect have to balance their beliefs against their commitments, against re-election prospects, against changes in the political landscape. Would we be wise to sit back and expect them to do what we need in the long term?

INTERNATIONAL CO-OPERATION

We have already seen one example – reducing marine emissions – where international co-operation is essential; without it, countries can put themselves at a disadvantage (typically economic) despite acting in the best interests of all. The UK is lucky; we belong to lots of clubs, each of which is keen to end humanity's dependence on polluting sources of energy and material. With our agreement these groups have made commitments[218] and provide support for keeping them:

> **United Nations** - in Paris 2015, 195 nations agreed to keep well below a 2°C warming limit. To meet the target the delegates made a host of agreements and decisions. These agreements and decisions are legally binding but do not include the emissions targets of individual countries. However, they do hold us all to a framework of co-operation and support that should drive reduced emissions and better mitigation for the kind of impacts we looked at earlier.

The **OECD** helps countries design environmental policies that are economically and environmentally effective.

The **IEA** works to ensure reliable, affordable, clean energy for its 29 members.

The **G7** countries have agreed to limit global warming to 2°C and have made a binding agreement to reduce greenhouse gas emissions (setting a 2050 target) and to decarbonise by the end of the century.

The **G20** works to set policy and share best practice in many arenas including energy efficiency, renewable energy, market transparency, and the phase-out of fossil fuel subsidies. China, which holds the G20 presidency in 2016, is also keen to promote circular and sustainable economies and halt the flow of plastic pollution.

Closer to home, the **EU** has agreed two binding agreements, as mentioned previously. In the first, the whole community decided to get 20% of its energy from renewable sources by 2020. The second ups the ante to 27% by 2030. We have also related supporting activities throughout this book, from taking freight off the road to building an energy grid linking Europe and North Africa.

---

218 Unfortunately, these agreements have not always played out as planned, eg Kyoto. You already know our role in this.

OTHER GOVERNMENTS

In setting our expectations of the support we should expect from government, it doesn't hurt to see what others are doing. Here are some examples from around the world:

Sweden's approach to taking fossil fuels out of its energy mix interests and inspires policy makers around the world. Some commentators compare the German *Energiewende* ('energy transition') with Sweden's energy policy. One big difference is in the way the two countries set their targets. Sweden put in place emissions targets and let the energy companies sort out the best technology. In Germany, the government selected technology, and then told the energy companies what to achieve. The two countries are in different phases of implementation: Germany has barely begun while Sweden set out to reduce its oil dependency in the Fifties.

In **California**, $86 million of the income from its cap and trade scheme will pay for solar panels on low-income homes.

In June 2015 the Obama administration extended and joined two nature sanctuaries on the **North Californian** coast, and banned offshore oil drilling within the new boundaries.

In 2007 the **Canadian Province of Ontario** passed a law to decommission all existing coal-fired power stations; it completed the phase-out of coal in 2014 and eliminated 'smog days', which had totalled 53 in 2005. The Ending Coal for Cleaner Air Act followed in 2015 and will promote a prosperous low-carbon economy, as well as recognise the impact of climate change on that corner of Canada.

Ontario is not alone; several cities, regions and countries have mandated decarbonisation. For example, **New York State** will get 50% of its electricity from renewables by 2050 as well as protect nuclear generation 'upstate'.

India plans to increase renewable capacity to 175GW; more starkly, it will also end fossil fuel subsidies.

Alberta – home of the Canadian tar sands – will implement a province-wide carbon tax and limit emissions from oil operations.

There are **38 carbon pricing schemes** in place or planned around the world. Together they cover about 40 countries, over 20 cities/regions and 12% of global emissions. There are two types of carbon pricing schemes. Direct taxes do what they say on the tin, and cap and trade schemes (CaT or ETS) give carbon dioxide producers an 'allowance' of carbon, which they emit or sell. Just in case you were wondering, here's a list: Finland carbon tax (1990), Poland carbon tax (1990), Sweden carbon tax (1991), Norway carbon tax (1991), Denmark carbon tax (1992), Latvia carbon tax (1995), Slovenia carbon tax (1996), Estonia carbon tax (2000), EU ETS (2005), Alberta SGER (2007), Switzerland ETS (2008), New Zealand ETS (2008), British Columbia carbon tax (2008), Switzerland carbon tax (2008), RGGI (Regional Greenhouse Gas Initiative, Northeastern United States and Eastern Canada) (2009), Ireland carbon tax (2010), Iceland carbon tax (2010), Tokyo CaT (2010), Saitama ETS (2011), Kyoto ETS (2011), California CaT (2012), Australia CPM (2012-2014), Japan carbon tax (2012), Québec CaT (2013), Kazakhstan ETS (2013), UK carbon price floor (2013), Shenzhen Pilot ETS (2013), Shanghai Pilot ETS (2013), Beijing Pilot ETS (2013), Tianjin Pilot ETS (2013), Guangdong Pilot ETS (2013), Hubei Pilot ETS (2014), Chongqing Pilot ETS (2014), France carbon tax (2014), Mexico carbon tax (2014), Korea ETS (2015), Portugal carbon tax (2015), South Africa carbon tax (2016) and Chile carbon tax (2017).

In conclusion

Deciding whether "To oil, or not to oil?" is scary. Moving away from oil needs more work than sitting tight does; indeed, waiting makes the inevitable change more difficult. Let's take some inspiration from the impossible-seeming goals we have already achieved:

> The **move from coal to oil** took most of us 70 years, but the British Navy accomplished the transition in a mere seven. This speed was due to the insight of Admiral Fisher and the determined First Lord of the Admiralty, Winston Churchill. Churchill took three years, from the start of his tenure at the Admiralty, to convince Parliament to increase Britain's naval might with new technologies including oil power. In the same period, Britain took

hold of 51% of Anglo-Persian Oil (which eventually became BP) to safeguard oil supply. By 1918 virtually the entire fleet ran on oil. The Navy achieved its goal despite the costs and difficulties of war and unproved technology (for example, an earlier venture to power the Navy with oil had ended in an explosion).

World War I saw a rapid advance in aircraft numbers and capability:

> *In 1914 the United States aeronautical industry employed 170 people, and the United States Army Signal Corps Aeronautical Division had eight active pilots. By the time peace broke out, the Army Air Service had grown to 197,000 officers and enlisted men, and American industry had produced 11,754 aircraft.*

> *The French had 260 aeroplanes before the war; during hostilities they built 68,000 and had 4,500 left at the end.*

> *Britain had 29 aeroplanes when the war started and 88 trained pilots. By the end of the war 54 types of fighter, reconnaissance and bomber aircraft had served in the forces that became the Royal Air Force on 1 November 1918.*

> *Before 1912 military leaders saw aircraft as little more than observation tools. Investment and use helped combat aircraft prove their value and become faster, easier to fly and more manoeuvrable. They also got weapons.*

After joining World War II, the United States **converted factories to make aircraft** and produced 300,000. The rest of the world made another 48,000.

In World War II, the Allies developed **penicillin** to support the D-Day landings. With concerted effort, the medicine grew from an aside in a paper written by Alexander Fleming to a mass-produced drug in a few years.

Much nearer in time and subject was the changeover from town to natural gas when the UK developed the North Sea gas fields. Between 1967 and 1977, the British Gas Council **converted every part of the gas infrastructure and every gas appliance** in the country.

Many of us are old enough to remember the furore that followed the discovery of the **Ozone Hole** over the Antarctic, widely announced in 1985. Ozone became the hot topic, with an almost prescient division of parties between those who 'believed' in the hole, and those who 'denied' its existence or repudiated the science that said we had to stop using certain chemicals. In May 2015 scientists from the University of Leeds published research showing how big the hole could have become if we hadn't acted, and the depletion that, most likely, would have affected Northern Europe (including the UK). Perhaps the most striking feature of the story was the non-story. We didn't stop having refrigeration, we didn't stop having aerosols, and we didn't halt the global economy; instead, we made an international agreement and acted on it – and now a whole generation has no idea how close we came to frying ourselves in ultraviolet radiation.[219]

In 1985, when Mikhail Gorbachev started the reforms of *glasnost* and *perestroika*, few would have predicted the **collapse of the Soviet Union** six years later. During the same period, the political environment created by Gorbachev blossomed into open borders between East and West Germany and the fall of the Berlin wall. Within a year, the two halves of Germany had reunited with a common government and currency.

In 1994, four years after Nelson Mandela left Robben Island, despite a mass of opposing political organisations and conflicts, the people of South Africa voted in a democratic election and **ended apartheid**.

The people of Kiruna in Northern Sweden faced 'a dystopian choice' in 2003. Do they let the largest iron ore mine in the world destroy their city, which would deprive the miners of a home? Do they close the mine, which would deprive the city of its biggest income? Or do they **move the city three kilometres** away to safety, create a Mine Park (so that no one has to live next to the industrial fence) and use the waste heat from the mine and processing to heat their new homes? Given the context, you should be plumping for the third option, as has Kiruna's town council.

---

219 A case of out of the fire, into the frying pan?

All these examples show that belief in change, robust plans and commitment make the biggest changes possible. These situations are only 'different' to our oily dilemma if we decide it is so. Throughout this book, you have seen the problem-solving inventiveness of a handful of people. Imagine what we could do if we all engaged in moving away from oil.[220] If you would like to see some pointers, you will find many countries' suggested plans for 'decarbonisation' in the Sources. Just remember: ambitious doesn't equal impossible. We make 90m cars and trucks every year; would turning some of that effort to wave power be impossible? We make 300m bottles of Champagne a year; could we use some of the energy to make solar cells? The replacement for Trident will cost the UK taxpayer £100bn over its lifetime; could we instead use a tiny fraction of that sum to bring permanently our old and vulnerable out of energy poverty?

So let's not forget how dozens of small actions can add up. We have used the phrase 'silver buckshot' a few times. If you think of the subjects we have covered, there is no silver bullet for any of them – much less a single replacement for fossil fuels. Instead, we are looking for silver buckshot: a *set* of alternatives to oil, each matching local needs. Turn back to Part One; you can probably reduce your personal oil consumption in each of the areas we covered with no (negative) impact on the quality of your life.[221]

Government and business are responsible for connecting the buckshot, and, here and there, they are doing so. From the example of first-generation biofuels driving up food prices, we know that oil, its uses and its alternatives are all related. Instead of saying 'stop', we have to say 'fix it'. In the biofuels example, 'fix it' means develop second-generation fuels – and you can see governments supporting new fuel technologies. We need an open debate to find the solutions with best potential, putting them in place quickly, mitigating the hazards and taking advantage of unexpected opportunities.

---

220 The idea that we can work our way through to a low-carbon future is gaining traction. Hopefully you have had a peek at the plans mentioned earlier in this chapter. You might also check out Global Apollo, Mission Innovation and the Breakthrough Energy Coalition. They all act on the premise that if we decide to make the move away from fossil fuels, we can be successful – very successful.

221 This can be as simple as catching and replacing a fallen button.

This book also occasionally mentions 'sustainability'. If you back off when you hear the word 'sustainable', don't. Let's skip the somewhat dull dictionary definition and turn to literature:

> *"Annual income twenty pounds, annual expenditure nineteen [pounds] nineteen [shillings] and six [pence], result happiness. Annual income twenty pounds, annual expenditure twenty pounds ought and six, result misery."*
>
> – Mr Micawber, in David Copperfield by Charles Dickens

Living within our means will not be straightforward. We need to expect some steps back to step forward. We need some chickens to come before the eggs. We will invest in technologies or economic theories that do not work well enough. Some people will seek to make a quick buck without delivering what they promise. Pressure groups will skew the arguments and forestall action out of self-interest. We will make compromises on the big and the small to protect the essential. No solution will be 'all things to all men'.

Oil presents a dilemma for each and every one of us. We cannot continue along traditional lines of party politics and self-interest; we need to act together if we are to reduce oil use.

As we reach the end of the story of oil, the message is:

> *"If necessity is the mother of invention, now is the time to invent our future."*

Close to the end

Some appendices

Appendix One – The car dilemma

The car dilemma illustrates some of the decisions we will have to make, using the car as an example. There are three parts; each makes a different argument.

Part One – affordability

While oil prices are low, you might decide to hang onto a less efficient car. If the volatile, fickle, unpredictable, undependable price of oil pushes up petrol prices or impacts your income, you might reverse that decision and look to buy a more efficient model. Every part of your car uses energy in production, and it features a lot of plastic. When oil prices rise, other prices rise, and new cars will cost more than they do today. When oil prices fall and affect your salary or pension, new cars are less affordable. Either way, the volatile, fickle, unpredictable, undependable price of oil will leave you less able to buy a more fuel-efficient car.

Part Two – new doesn't mean less oil

According to UNESCO, making a car uses 20,000 megajoules (MJ) of energy – about 545 litres of oil. My small old hatchback consumed eight litres of petrol for every 100km driven. The new version of the same car uses 5.5 litres of petrol to cover the same distance. To recover the energy used to make a new car, I would have to drive more than 19,000km (12,000 miles); only after that point will the new car 'save' any oil.

This simple calculation isn't accurate, primarily because it incorrectly assumes the energy content of oil and petrol are equal. Neither does it factor in the inevitable degradation in fuel efficiency over time. But more accuracy would not improve the answer because making stuff uses oil.

Part three – getting started

If we are going to change from petrol and diesel to non-fossil fuels, we have a chicken-or-egg problem. Increasing the renewable content of fuel will damage some cars and force their owners to take otherwise-working motors off the road or face bills for repairs and upgrades. The oil burnt to make new cars or parts will offset the oil benefits of increasing the alcohol content. (When made from foodstuffs, alcohol uses a lot of energy. Producing alcohol from maize[222] uses almost as much energy as it later releases, and the focus on maize-based ethanol has detracted from investment in second-generation biofuels.) So, starting with the fuel could seem foolish, until you consider changing the cars first. Developing cars that can tolerate alcohol and take full advantage of it costs money – money that car companies would not spend without a proven market. Second-generation fuel producers need the same confidence to invest in new technologies. We all need a resilient infrastructure to deliver biofuels to consumers.

Which would you introduce first: the fuel or the cars? How would you manage the older vehicles still on the road? How would you encourage investment in new fuels?

---

222 Americans call maize 'corn'. In England, 'corn' often refers to wheat, though it can be used of any grain (hence, a little strangely, peppercorn). In Scotland it refers to oats. The Australians and New Zealanders side with the Americans on this one.

Appendix Two – How technology advances

Why does technology advance?

Technology advances because someone somewhere asks a question:

Problem solving – To address the problem of anaemia in Cambodia, Christopher Charles asked: "How can I get more iron into the diet of local people?" The result is a small iron fish. Shaping the iron as a fish encourages people to put it in their cooking pot to bring luck. Once there, it adds iron to the food.

New ideas – When they developed the capacitive touch screen, the research team at the University of Delaware were answering the question: "Can I control a device entirely through its screen?" (I speculate on the exact form of the question.)

Discoveries – By asking "What happens if I turn a magnet inside a metal coil?" Faraday proved his theory of electromagnetic induction.

Who pays for technological advances?

Few industries or businesses can fund the research they need to keep, or gain, a competitive advantage. Therefore, many rely on government support. This support comes in different ways:

- Direct grants;
- Access to university or government research;
- Commitment to creating an attractive economic environment;
- Government projects; eg military and space.

In the UK, companies can seek support from local government, national government and the European Union.

Governments end up funding research because it is risky; fundamental research makes no guarantee of financial return, so banks and other financial institutions are unwilling to support it. The Sources gives links to an article, report and presentation showing how the US government has paid universities and companies to develop technologies from interchangeable parts to key components for the iPhone.

Who buys new technology?

You may recognise the term 'early adopter' and probably know someone who is always the first to get their hands on new technology. The five types of technology adopter are:

- Innovators;
- Early adopters;
- Early majority;
- Late majority;
- Laggards.

Technologies mature

Before they reach us, technologies go through a vast array of development stages, from proving scientific principles to setting up volume production. Just think back to the dates in "Taking oil my energy"; the first AC power station came 50 years after Faraday's demonstration. Similarly, inventors and early adopters would barely recognise modern technologies. Everything we use has gone through generations of improvements and updates. Here are a few examples you might find interesting.

CHRISTMAS TREE LIGHTS – BURNING DOWN THE HOUSE

In my home, we use our old Christmas trees as kindling, as the dry needles burn well. So I was surprised to learn the first lights on Christmas trees were candles. And even after Edison put incandescent bulbs on trees, the heat from the lamps, combined with ideal burning fodder, made for evenings of light and heat for all the wrong reasons.

TRAINS – STEAM TO BULLET

Horse-drawn carriages have run on tracks or rails since Roman times, but the real dawn of rail travel came with a rapid series of development building on Trevithick's steam-driven Pen-y-Darren (1804). These efforts resulted in the first freight railway in 1825 (though horses continued to pull passengers until 1833). Queen Victoria was a little apprehensive about travelling at the dizzying speed of 40mph; can you imagine telling her that passengers of all classes can expect comfort and speeds of 311mph on Japan's newest bullet train? When you think about its forebears, Puffing Billy, Catch Me Who Can and The Rocket, you can only wonder why this technical marvel has the name 'L0 Series'.

WINDOWS – GLASSES AND THE LAGER STORY

We build solar panels into windows. We manipulate glass molecules to stop heat loss or gain. We make enormous sheets of glass to adorn skyscrapers and the houses on TV shows. Glass is cheap, and we take it for granted; however, historically, glass was the preserve of the wealthy. Glass windows were a marvel; glass beads changed hands at prices that could keep a whole family for a year; glass-fronted cabinets displayed glass bowls, plates and drinking vessels.

Then technology lent a hand. Even though the Romans made glass, the collapse of the Empire meant that British windows went unadorned for centuries (excepting churches and other exalted public spaces). Glass manufacturing rebuilt slowly and with expertise from Italy. In the 17th century, Ravenscroft discovered that adding lead oxide to melted glass improved its properties (hence the name lead crystal). Methods of making glass sheets grew and developed from various blown methods (like crown or bulls-eye glass). Like metal, glass can be cast, polished and rolled. These three methods remained the mainstay of sheet-glass making until Pilkington developed float glass, which went into production in 1960.

We use glass for many purposes and have given its name to one object in particular – the drinking glass. Until the 19th century, beer came in only pewter or ceramic mugs. One school of thought has it that glassware (which was cheaper) showed drinkers how murky their favourite tipple was and opened their arms to filtered beer, especially lager. Alternatively, some say the brewers of clear beer promoted glass to set off their product to its best advantage. Either way, innovation in glass is linked to the decline of traditional brewing.

BEER – FROM ALE TO MASS- TO MICRO-BREWING

Beer has a long history; in medieval times alewives made beer alongside bread. Over the centuries, brewing moved into dedicated buildings and adapted to local conditions. Then came the Burton brewery style. Bass (owner of the first UK trademark) made a new style of beer thanks to the character of the water they drew from local wells. This beer was a Pale Ale, but it was more astringent and tasted more strongly of hops than previous Pale Ales. To match the refreshing brew, other breweries began to take their water through a process called Burtonisation.

The near universal adoption of Burton style beer started the scientific, automated approach to beer making, and by the end of the 20th century many breweries closed or were consumed by international giants. Luckily, beer has shown us that we can reserve a move to large-scale production. Indeed, even as prices rise, crafted beers from microbreweries show us 'where there's a will, there's a way'.

Thinking outside the box

What a horrible clichéd phrase. Let's explore it (and explain ourselves) with some examples of technologies which are approaching things differently:

In the UK we flush two billion litres of clean water down the toilet each day. Thinking inside the box led us to low-flush toilets, toilet 'hogs', and only flushing when necessary. Thinking outside the box leads us to ask 'why do we use water?' The answer is 'because the Romans flushed'. The Romans did a lot for us, but wasting energy and materials just because *they* washed away waste seems a little short-sighted. So researchers are exploring **the composting toilet**. Not only do dry loos save water, but they also produce electricity and fertilisers. Please do not poo-poo the idea.

Another water story swims at the opposite end of the supply chain. Taking the salt out of water in drought-ridden countries can be an expensive affair. The modern technology is filtration, which is material- and energy-expensive. But Californian engineers have turned back to **an age-old method of taking salt and other minerals out of water**: They use distillation. Rather than heating the water with fossil fuels, they run their process with solar energy. This 'back to basics' approach uses a plentiful Californian resource to supply a scarce one.

Brazil is hot and dry: not ideal conditions for a country that relies on agriculture as a mainstay of its economy and the primary source of fuel. Farmer and, one might say, visionary Herbert Bartz suffered a catastrophic flood in 1971 and lost half his crops and topsoil. He looked for a way of conserving his soil and turned to a fledgeling technique called **'no till'**. In this method of farming, old crops are left in the ground; this preserves water and holds on to organic matter and minerals. Herbert was so successful that 75% of Brazilian farmers have stopped tilling their land. No till wasn't an overnight transition; Herbert worked hard to tailor the new techniques to hot, arid Brazil.

'Out of the box' takes a special way of thinking. Keep an eye open and you will find many people have it and are tackling decarbonisation in novel and surprising ways.

> *"If you always do what you always did, you will always get what you always got."*
>
> *– Attributed to Albert Einstein, Henry Ford and many others*

Appendix Three – Electrical terms and units

Electricity is hard to understand. Since static electricity and electricity with a current are not two types of the same thing, the following does not apply to static electricity.

The energy stored in batteries or which comes out of the wall makes bits of atoms move. As these bits jog around, they pass on their energy to the next atom in line. When the energy passes down a chain of atoms – say, in a wire – we can use it. We call that energy electricity. Electricity has lots of characteristics:

> **Current** is like flow; it (kinda) tells us how much energy passes through a material – a bit like knowing how much water moves downstream in a river.

> **Voltage** tells us how much the energy 'wants' to move. In our river analogy, it is like the difference in height between the source of the river and its end.

> **Power**. If you have a river with lots of water (current) and it falls a long way (voltage), then you will have lots of power. Imagine the difference between standing under your shower and standing at the bottom of Niagara Falls (other waterfalls are available).

> **Energy**. If your waterfall runs for a very short period it will have less energy than a waterfall that runs for hours. Think about putting a watermill at the bottom of the waterfall: if it turns for a couple of seconds, you will grind much less wheat than you would if the wheel turned for the best part of a day.

In *The VFUU Price of Oil* we talk about electricity using four units (there are many more):

> **amps** – the unit of current (A) (named after André-Marie Ampère, a French mathematician and physicist);

> **volts** – the unit of voltage (V) (named after Alessandro Volta, an Italian physicist and chemist);

> **watts** – the unit of power (W) (named after James Watt, a Scottish mechanical engineer and chemist);

> **watt hours** – the unit of energy (Wh). There is also an energy unit called the Joule (named after James Prescott Joule, an English chemist and physicist).

But these units are tiny. Think about AA batteries: they deliver 1.5V. One watt is the power rating for a single LED; in one hour that LED would use one watt-hour (1Wh). More often we talk about much larger quantities of electricity – thousands, millions, billions, and trillions of times more. Rather than write out an enormous number with lots of zeros, we use a prefix (a little word in front of another word). The International System of Units (SI) defines the prefixes as follows:

- A thousand – three zeros – is *kilo-* (think kilometre)

- A million – six zeros – is *mega-* (think megabyte per second, the speed or bandwidth of your Internet connection)

- A billion – nine zeros – is *giga-* (think gigabyte: about two movies' worth of data)

- A trillion – twelve zeros – is *tera-* (think a large hard drive).

(IT geeks – I am using the SI definitions here, not the binary definitions; everyone else: don't worry about it.)

To make things easier to read, we usually shorten the names. So you will find a gigawatt called a GW. In each section, the first time the abbreviation is used, it comes with the full name of the unit to help you remember what it means.

Appendix Four – The chemistry of carbon (organic chemistry)

Carbon is jolly useful stuff. We have relied on its unique properties for millennia. It has traditional uses, like graphite, and modern forms, like graphene. (Quick carbon fun facts: The term 'black market' comes from the illegal trade in graphite in the 19th century.) Graphene is a single layer of graphite. The first graphene created measured 0.33 nanometres; that is, 0.00000033mm – or, to put it another way, it is a million times thinner than a human hair.[223]

Carbon is one of 118 naturally-occurring and synthetic chemical elements. Other familiar elements are oxygen, hydrogen, iron, copper and thorium. Atoms are the smallest, indivisible pieces of elements. When we split atoms, they are no longer elements.

Each atom consists of protons and electrons. Excepting hydrogen, they all have neutrons too. Protons and neutrons sit in the centre (nucleus) of an atom. Electrons orbit the nucleus. You can think of the electrons circling the nucleus like the planets orbiting the Sun.[224]

The number of protons defines the element. All hydrogen atoms have one proton; all carbon atoms have six protons. Some elements have multiple varieties or 'isotopes'. Isotopes have nuclei with different numbers of neutrons. For example, the most common form of hydrogen contains no neutrons; chemists call it protium. Deuterium is the hydrogen isotope with one neutron, and tritium has two. Deuterium is relatively common on Earth: in seawater, for every million hydrogen atoms, about 150 will be deuterium. Its quantities vary elsewhere on Earth and in the universe. Deuterium is harmless. On the other hand, tritium is radioactive and, because there are only trace amounts on Earth, we have to make it.

Atoms contain equal numbers of electrons and protons, but some quantities of electrons are special. Helium has two protons and two electrons. Two is a *stable* number of electrons; so helium atoms do not

---

223 Before we start, I would like to state that just the few items in this appendix fill chemistry books. I have made simplifications to make it understandable and short. When things are simplified, some accuracy, or precision, is necessarily lost. So please take this introduction as a way of showing how splendidly versatile carbon is, and nothing more.

224 This is the easy way to think of electrons, but it is not precise enough to fully describe their behaviour. If you want that degree of precision, refer to the Sources.

tend to combine with other atoms. The other stable numbers are 10, 18, 36, and 54. We call the elements which have those numbers of protons and electrons the Noble Gases because they do not mix. Noble Gases with more protons and electrons are radioactive.

Elements that don't have a stable number of electrons like to mix or react. If they are one or two electrons short of a stable number, then they take electrons from other atoms. Oxygen has eight electrons; it 'wants' ten, and takes the missing two from metals. When it does this to iron, the iron rusts.

Atoms which are one or two electrons richer than a stable number give them away. Sodium has eleven electrons and 'wants' to shed one. If you mix sodium and oxygen, they react. Oxygen yields energy as it gains electrons from sodium and sodium gives out energy as it loses them. We sense the energy as an orange flame.[225]

Oxygen is two electrons short, but sodium only has one spare. So, each atom of oxygen (chemical symbol O) combines with two sodium atoms (chemical symbol Na) to make sodium oxide. Chemists write all that down like this:

$4Na + O_2 \rightarrow 2Na_2O$

When a metal (sodium) and a non-metal (oxygen) get together, they form ionic bonds. The name reflects the way atoms turn into 'ions' by gaining or losing electrons.

When two non-metals bond, they share electrons. The technical name for sharing electrons is a 'covalent' bond. For every pair of electrons shared, there is a bond; for example, sharing three pairs of electrons creates three bonds, called a triple bond. Carbon has four electrons to share, shown as dots in the diagram.

---

225 As a tiny aside, losing electrons (to oxygen) is called *oxidation*; gaining them (from sodium) is *reduction*. At school I learnt the handy mnemonic 'oilrig' – oxidation is loss, reduction is gain.

Figure 30: Carbon

When carbon (the slightly smaller atom with dots for electrons) reacts with oxygen (the slighter larger atom with stars for electrons), it has two choices. It can share two electrons with one atom of oxygen to make carbon monoxide. In total the atoms share six electrons (a triple covalent bond).

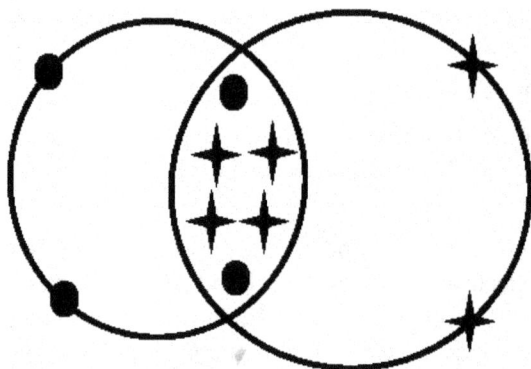

Figure 29: Carbon monoxide

Alternatively, an atom of carbon can share two electrons with each of two atoms of oxygen to make carbon dioxide. Each oxygen atom (stars) shares two electrons with the carbon (dots) –making two double covalent bonds.

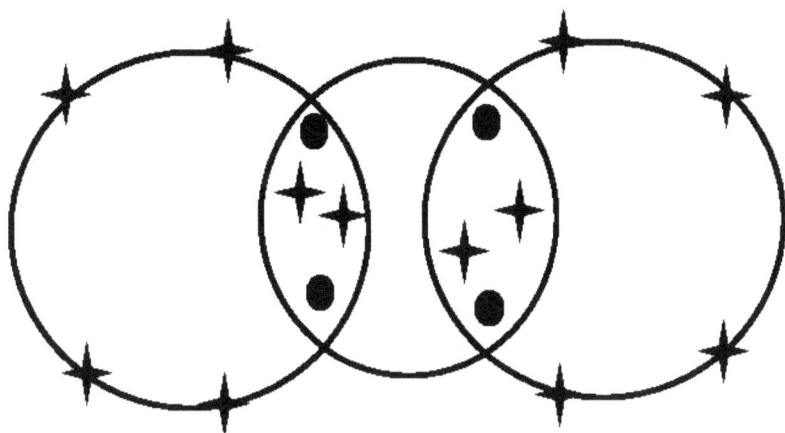

Figure 31: Carbon dioxide

Don't worry about the chemistry. Just note the flexibility of carbon; very few other elements can make so many compounds. When our flexible friend bonds with hydrogen, the possibilities are endless.

Let's start with the simplest carbon-hydrogen molecule. It is methane and has the chemical formula CH_4 – in other words, methane is four hydrogen atoms attached to one carbon atom. Each hydrogen atom (the small atom with triangular electrons) shares one electron with the carbon (dots) – making four covalent bonds. The methane molecule is a tetrahedron – a four-sided pyramid – with the carbon sitting in the centre.

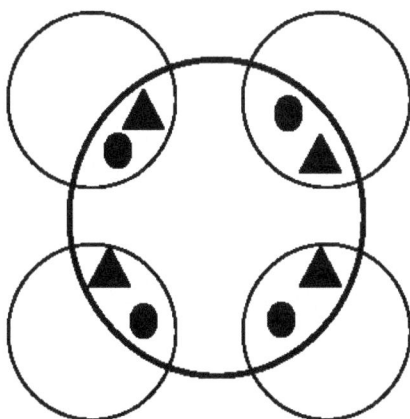

Figure 32: Methane

Flexible carbon can also combine with other carbon atoms. In ethane, two carbon atoms form a single covalent bond. They also make covalent bonds with three hydrogen atoms each.

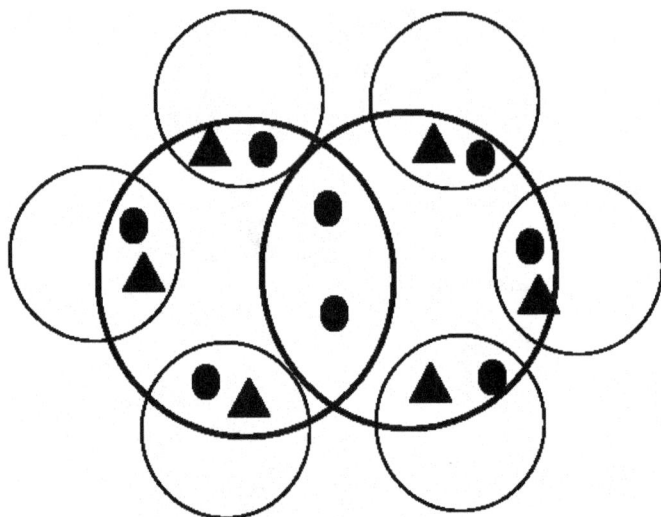

Figure 33: Ethane

Given the high number of possibilities, we find hydrocarbons with chemical formulas like those in figure 34. How much easier to say *I need my morning caffeine*, than *line me up with some of that excellent* $C_8H_{10}N_4O_2$? But there are so many hydrocarbons that they cannot all have common names (in the table above there are 11 common names and one scientific name). The scientific names for hydrocarbons follow rules so you can deduce a fair amount about them from their names. The names tell us how carbon atoms connect with each other and with hydrogen. They show each hydrocarbon's properties.

CH_3COCH_3, or $(CH_3)_2CO$ (acetone)	$C_{10}H_{18}O$ (linalool)
$C_5H_8O_2$ (angelic acid)	$(C_2F_4)_n$ (Teflon, admittedly not a hydrocarbon, but still making use of carbon's flexibility
C_6H_6 (benzene)	$C_2H_6Cl_2Si$ (dichloromethylsilane)
$C_8H_{10}N_4O_2$ (caffeine)	$C_{12}H_{22}O_{11}$ (maltose)
$C_{13}H_{18}O_2$ (ibuprofen)	CH_2O (formaldehyde)
$C_7H_6O_3$ (salicylic acid)	$C_{35}H_{49}O_{29}$ (xanthan gum)

Figure 34: Some common hydrocarbons

Hydrocarbon families

ALKANES

The gases we use as fuels all end in '-ane'. These hydrocarbons are part of a family called the alkanes. Each carbon pair has a single covalent bond. They are not reactive, but they burn well in oxygen to give carbon dioxide, water and heat. If they burn without enough oxygen, they make carbon monoxide.

ALKENES

The chemicals that make polymers have names that end in '-ene' (who said organic chemistry was hard?). The simplest, most familiar alkenes have one double covalent bond. The bond is key to their plastic properties.

303

CYCLOALKANES

We have not met any of these hydrocarbons, but they are worth mentioning as the carbon forms rings – the smallest has three carbon atoms. These ringed hydrocarbons have only single covalent bonds.

ARENES – OR AROMATIC HYDROCARBONS

Benzene is the most familiar and simplest arene. It is a ring of six carbon atoms connected with alternate single and double covalent bonds.

ALCOHOLS

Alcohols contain an oxygen atom and are an essential product of fossil fuels.

Now we know what the endings '-ene' and '-ane' mean, let's think about the start of hydrocarbon names. Each prefix below refers to the number of carbon atoms in the molecule. Note: methane is the only hydrocarbon with just one carbon atom.

1. Meth-
2. Eth-
3. Prop-
4. But-
5. Pent-
6. Hex-
7. Hept-
8. Oct-
9. Non-
10. Dec-

Carbon dating

Carbon dating has little to do with oil but depends on the science of carbon. Carbon has six protons; this is what makes it carbon. The most common form of carbon, Carbon12 (C12), has six neutrons. It is stable. Carbon13 (C13) has seven neutrons and is stable. Carbon14 (C14) has eight neutrons and is radioactive.

When cosmic rays hit nitrogen in the atmosphere, they convert the nitrogen into C14 and hydrogen. The C14 turns back into nitrogen over time. Half of any amount of C14 decays into nitrogen in 5,730 years (this period is its *half-life*).

By a curious coincidence, C14 forms at roughly the same rate that it decays. So the Earth's ratio of C14 and C12 has remained fairly constant for a long time. There is one C14 atom for every trillion C12 atoms in the air, the sea, and living organisms. But as soon as we die, we stop taking in carbon, but the C14 in our bodies continues to decay. After 5,730 years, the ratio in our remains will be one C14 atom for every two trillion C12 atoms.

Archaeologists use this knowledge to estimate the age of dead organic material. Fittingly for our story, no fossil fuels contain C14 – they are too old. When we burn them, they release only C12 (and some Carbon13, but that does not count). In the fossil-fuel age, we have changed the proportion of C14 in the atmosphere, the sea and every living organism. Similarly, but to a lesser extent, C14 production rates are influenced by the variation in solar activity over a 9,000-year cycle. Today's scientists account for the latter. Their successors will adjust for our fossil fuel habits.

Appendix Five – Is the UK's shale gas all it's cracked up to be?

The gas industry, the government and the media tell us shale gas extraction in the UK will:

- Fuel the UK for several decades;
- Lead to economic growth;
- Reduce carbon emissions (by replacing other fossil fuels);
- Provide energy independence for the UK;
- Reduce gas prices (we won't get into this – just ask yourself whether our gas prices are likely to be lower than those in the rest of the world.)

Here we will examine the assertions made for the UK's Bowland-Hodder gas reserves. To test the accuracy of these claims, we use mathematical models of the extraction and use of gas in the UK under three sets of circumstances (scenarios).

In all three scenarios, we extract 10% of 1,329 trillion cubic feet of gas over the productive life of the Bowland-Hodder reserves. 10% is the number the press homed in on, so it is the one used here. The big number is the best estimate of gas in place made by the British Geological Survey in 2013. The table below shows the full range of their estimate. You read it like this: there is a 90% probability that the Bowland-Hodder reserves hold 822 trillion cubic feet ('tcf') of natural gas.

90% probability	50% probability	10% probability
822 tcf	1329 tcf	2281 tcf

Figure 35: The BGS's estimate for gas in place in the Bowland-Hodder reserves

For each scenario, read the assumptions made and decide whether you think they are correct. If you disagree with them, then estimate how your assumptions affect the result; for example, lower production rates would make the gas last longer, but we would have to import more annually. If you like, make your own calculations and charts.

The UK's shale gas scenario one

This scenario assumes that Bowland-Hodder builds production to eliminate the need for imports and accommodate the drop in conventional production.

ASSUMPTIONS

- The UK's shale gas comes online in 2019 and ramps up to full production in 2029.

- Existing gas production (chiefly North Sea) continues to fall an average of 8% per year (2014 bucked this trend but at the time of writing there is no certainty this can be maintained).

- No economic growth; therefore, consumption remains at 2014 levels.

- Natural gas does not displace coal for the production of electricity.

RESULTS

Scenario one - meeting our current needs

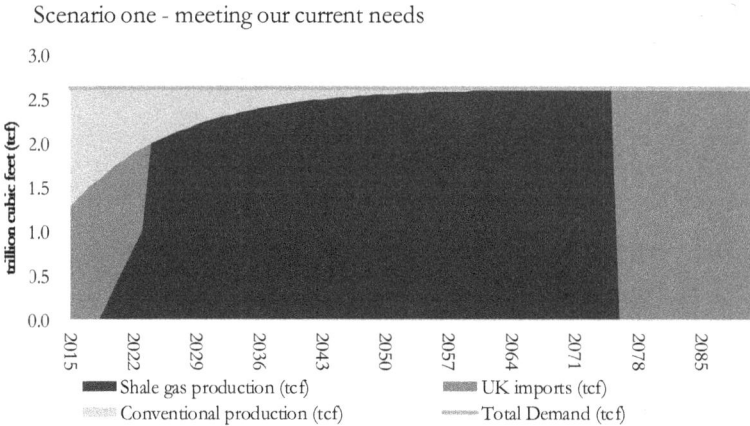

Figure 36: The UK's shale gas scenario one

- No energy independence; imports are eliminated, but then grow to match the decline in conventional gas production.

- Carbon dioxide emissions do not reduce because we will continue to burn gas, coal and oil at the same rates.

- The gas runs out in 2076.

The UK's shale gas scenario two

We build in economic growth and hold imports at 2014 levels.

ASSUMPTIONS

  ☥ The UK's shale gas comes online in 2019 and immediately grows to meet demand.

  ☥ As soon as the gas comes online the UK economy grows, leading to a 5% annual growth in gas consumption (this is the average growth in gas consumption from 1987 to 2013).

  ☥ Existing gas production (chiefly North Sea) continues to fall an average of 8% a year (2014 bucked this trend, but at the time of writing there is no certainty this will be maintained).

  ☥ The economy holds static when the gas runs out, and imports increase to meet demand.

RESULTS

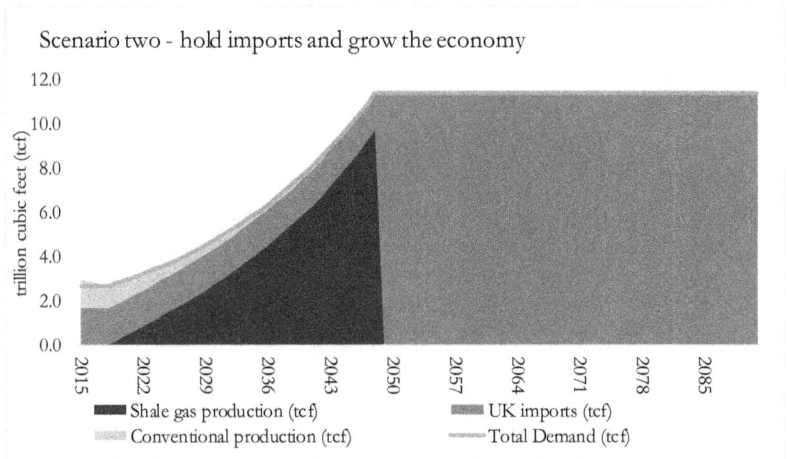

Figure 37: The UK's shale gas scenario two

  ☥ No energy independence; at best, imports hold while the gas lasts.

  ☥ The economy grows.

  ☥ Carbon dioxide levels rise because we increase gas consumption (and coal and oil).

  ☥ Extraction exceeds Texan (2014) levels in 2036.

  ☥ The gas runs out in 2049.

The UK's shale gas scenario three

We take scenario two and replace coal-fired electricity production with natural gas.

ASSUMPTIONS

As scenario two, plus:

- ꭥ Natural gas replaces all coal burnt for power generation (not industry), which starts at 2014 levels and increases by 5% a year to match the growth in natural gas consumption when the gas comes online.

RESULTS

Scenario three - hold imports, grow gas consumption, replace some coal

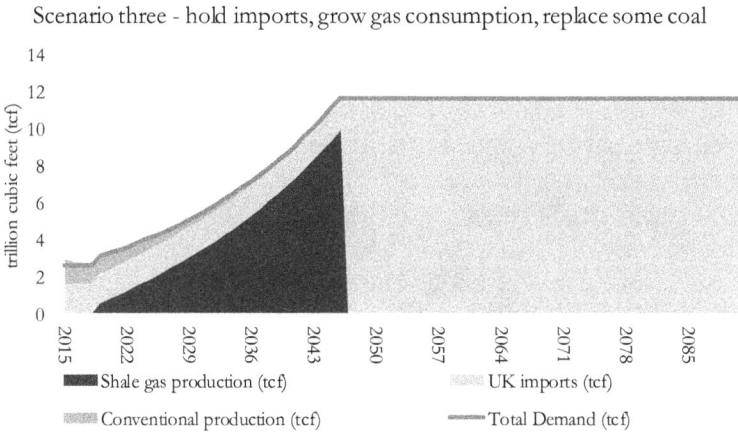

Figure 38: The UK's shale gas scenario three

- ꭥ No energy independence; at best, imports decrease while the gas lasts.
- ꭥ Economy grows.
- ꭥ Compared with scenario two, carbon dioxide emissions decrease on the back of reduced coal use for electricity generation.
- ꭥ Extraction exceeds Texan (2014) levels in 2033.
- ꭥ The gas runs out in 2047.

These scenarios show that the most utopian claims for shale gas are clearly ludicrous: unless we do not expect the impossible of the gas industry, shale gas cannot simultaneously fulfil the claims of energy independence, reduced carbon emissions and economic growth. Even if the gas industry should perform a miracle, the gas will last less than three decades. What's more, we will have increased (by over four times) the amount of gas we use, and will have to deal with the political, economic and environmental impacts just as the economy slows down. (If you are wondering why the charts look a bit funny between 2015 and 2017, it is because exports are not in the models.) Of course, individual operations could be extremely successful, but this analysis shows that we cannot rely on UK shale gas as a fuel and to deliver the economic benefits espoused by its advocates.

The logistics of UK shale gas exploitation

The scenarios above are not only crazy in the ramp-up and levels of production; by assuming we can extract 10% of the gas in place we end up with a somewhat silly logistical situation. The number of wells and pads given in each scenario show the amount of space needed to extract UK shale gas. They show why finding more gas will not help. Assuming we aim to extract 10% of the Bowland-Hodder reserves, we will run out of space before we achieve our goal. To calculate how many wells and pads we might need in the Bowland-Hodder, we have to look at quite a bit of data. First we need to know how much gas a well can produce (its ultimate recovery rate, or URR). The answer, in billion cubic feet ('bcf' here), depends on who you talk to (or look up on the Internet); the following is the range of answers and sources used for this appendix.

Source	UoT max	UoT min	UKECR	Navigant	'Industry' figure	EIA
Extraction per well (bcf)	4.3	0.4	2.3	2.1	2.65	1.4

Figure 39: Extraction per well

Wells sit on 'pads', and all operators put multiple wells on pads. Cuadrilla says it can fit 32 wells on a single pad and there are certainly examples of operators getting over 50 wells on a pad, but we assume a more modest 20. To extract 10% of the likely gas in place in Bowland-Hodder will require the following number of wells and pads:

	Number required to extract ten per cent of gas in place
Minimum number of wells	30,907
Maximum number of wells	332,250
Minimum number of pads	1,545
Maximum number of pads	16,613

Figure 40: Number of wells and pads required to extract 10% of gas in place

These figures reflect the total life of the shale play so they will be spread out over time. If we consider the peak production in each of three scenarios, then we can determine the largest area of the Bowland-Hodder field that will be taken up by exploitation at any one time.

SPACE REQUIREMENTS

Scenario and extraction at peak	Scenario one 2.6 tcf	Scenario two 9.6 tcf	Scenario three 9.9 tcf
Smallest estimate of number of wells at peak	611	2,253	2,317
Largest estimate of number of wells at peak	6,572	24,215	24,905
Smallest estimate of number of pads at peak	31	113	116
Largest estimate of number of pads at peak	329	1,211	1,245

Figure 41: Peak number of wells and pads

How much space will these wells and pads take up? Each pad is about the size of two football pitches during drilling, and about half the land is restored after the drilling phase. Here are the number of square kilometres that we will have to cover in concrete to achieve the UK's shale gas aspirations, and some comparisons to help understand those figures:

	UoT (2013) max	UoT (2013) min	UKECR	Navigant	'Industry' figure	EIA
Wells for the entire operation (132.9 tcf)	30,907	332,250	57,719	62,722	50,151	94,929
Pads assuming 20 wells each	1,545	16,613	2,886	3,136	2,508	4,746
Space (km2) at 0.014 km2 per pad	22	233	40	44	35	66
Equivalent sized towns/cities	Camden (London) 21.79	Kettering (Northants)	Reading (Berks) 40.40	Eastbourne (E.Sussex) 44.16	Blackpool (Lancs) 34.92	Chesterfield (Derbys) 66.04

Figure 42: The space needed to extract 10% of the gas

Thinking about population density only adds to the logistical woes. The population density of the four major counties of the Barnett Shale are:

- Tarrant – 809 people per km²
- Denton – 498 people per km²
- Johnson – 80 people per km²
- Wise – 25 people per km²

Lancashire (in one of the UK shale gas sweet spots) has 1,419 people per square kilometre.

So, we need an obscene number of wells, pads and tons of concrete, and cannot physically get to all the Bowland-Hodder shale because people live on top of it. Any one of these factors makes that 10% recovery rate seem ambitious.

Appendix Six – What is climate change, and should we act to prevent it?

'Climate change' has become shorthand for the effects of releasing greenhouse gases into the Earth's atmosphere where they trap the heat of the Sun. We need these gases else the Sun's energy would bypass the Earth. However, as the volume of greenhouse gases in the atmosphere increases, the planet and its gaseous duvet quickly grow warmer. (The actual temperature is slightly less important than the rate of change, which is now quicker than most plants and animals can adapt to.)

But focusing on warming sets false expectations. Our weather patterns are complex and respond in unexpected ways to higher temperatures; for example, some areas experience cooling. Nor do temperatures and greenhouse gases have a straightforward relationship. Instead of, say, air temperatures rising in line with carbon dioxide, they may grow faster. If you are thinking *ah, but temperature rises have paused, plateaued or stalled*, check the article in the Sources. While underlying patterns of climate change are natural (think: ice ages) we shall see in a second that scientists are certain the patterns are changing and that greenhouse gases are the cause.

What are the greenhouse gases?

The four most common are:

> **Water vapour**, which enters the atmosphere in prodigious quantities and leaves again through rain. As the atmosphere warms, it can hold more water (which is why hot countries tend to be more humid than cold ones), but this does not have a significant impact on climate change as it quickly rains out again. To influence temperatures, water needs to be in the upper atmosphere (the troposphere) to accumulate. We add water to the lower atmosphere (the stratosphere) whenever we burn fossil fuels, and some of it does move upwards from the stratosphere, but (again) most of it rains out. Therefore, our greatest impact is through powered flight, which delivers water directly to the troposphere. The global warming effect of water and other particles from flying is nearly the same again as its carbon emissions and is called 'radiative forcing'.

Carbon dioxide is the most prolific of the remaining greenhouse gases. For this reason, the impact of the other greenhouse gases is always compared to it. Scientists use a number called the global warming potential (GWP) to describe this comparison, giving the GWP of CO_2 a value of one.

Our old friend **methane** is the second most common greenhouse gas if we discount water vapour, and has a GWP of between 28 and 34.

Nitrous oxide is no laughing matter and, excluding water vapour, ranks at number three in volume. People account for around 30% of N_2O emissions (the rest comes from the soil). Its GWP is 265-298 and it destroys ozone. Do not confuse it with other oxides of nitrogen, which merely make smog.

We do not need to be climate scientists to ponder the validity of climate change. You don't have to measure changes, create theories or test them. Whether we should act to minimise our effect on the environment depends on these three things:

What the scientists say

When scientists discover something, they tell the world in a paper. The most credible papers are those that have been reviewed by other scientists, aka their peers. Journals publish the papers and arrange the peer reviews. The reviewers check that the researchers used robust methods, assessing things like the design of an experiment or whether data have been acquired from reliable sources. Journals will not publish research that fails to meet the appropriate standards.

Peer-reviewed papers give rise to a strange form of scientific assessment - metadata. Metadata are data that describes data. Scientists use metadata to understand how published papers agree (or not). (Metadata in computer science describes data too, but are employed in a different way). Metadata about climate papers show that 97% of published climate scientists have found evidence to support climate change theories.

That doesn't mean scientists agree about every detail – would you expect otherwise? But the study does show that 97% of scientists who study climate change believe increasing levels of greenhouse gases warm our atmosphere and that that warming changes our weather patterns.

If you think 97% is not good enough, take heed of some other scientific consensuses. 97% of scientists believe in evolution, even though scientific data has been used to argue a biblical, rather than astronomical, age for the Earth. As recently as 1999 a scientist published a paper to demonstrate smoking does not cause cancer. Are you willing to take up smoking because some data shows it's safe?

> *"Scientific consensus is the collective judgment, position, and opinion of the community of scientists in a particular field of study. Consensus implies general agreement, though not necessarily unanimity."*

> *– Wikipedia*

The United Nations International Panel on Climate Change (IPCC)

The IPCC considers the science, impact and possible mitigations of climate change. Its name and activities suggest that the panel has made up its mind. However, working parties consisting of experts in many fields assess all the evidence and latest science, and report in three subject areas:

Working Group	Number of Authors	Number of countries	Number of comments
Working Group I The physical science base	259	39	54,677
Working Group II Impacts, adaptations and	308	70	50,492
Working Group III Mitigation of climate change	235	58	38,296

Figure 43: IPCC assessment and review statistics

The IPCC issued its first assessment in 1990. It released the fifth assessment in stages, from November 2013 to October 2014. Each working party considers the science in their area of expertise; eg sea level rises. They review the previous assessments along with new evidence. They write a draft paper, then open their work to comments and evaluate each one. They also prepare an overarching report and take it through the same stages of review.

Your perception of risk

If you look at the IPCC charts showing temperature changes; if you look at photographs showing rapid changes in landscapes; if you read about the decline and extinction of species[226] – you may ask yourself whether we want to address the possibility (risk) that greenhouse gases are the cause.

If you have a choice between energy that may cause climate change and sources of energy that do not, perhaps you could hedge your bets and select the latter. To do so is to mitigate the risk. Energy and climate change experts have shown the cost of mitigating against climate change is lower than the estimated costs of climate change. Indeed mitigation can bring financial benefits greater than the cost.

---

226 52% of wildlife lost in 40 years.

Appendix Seven – The other fossil fuels

Fossil Fuel primer

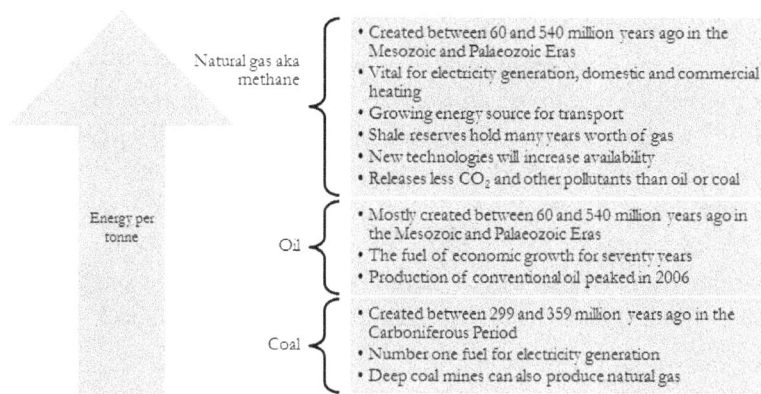

Natural gas aka methane
- Created between 60 and 540 million years ago in the Mesozoic and Palaeozoic Eras
- Vital for electricity generation, domestic and commercial heating
- Growing energy source for transport
- Shale reserves hold many years worth of gas
- New technologies will increase availability
- Releases less CO_2 and other pollutants than oil or coal

Oil
- Mostly created between 60 and 540 million years ago in the Mesozoic and Palaeozoic Eras
- The fuel of economic growth for seventy years
- Production of conventional oil peaked in 2006

Coal
- Created between 299 and 359 million years ago in the Carboniferous Period
- Number one fuel for electricity generation
- Deep coal mines can also produce natural gas

Energy per tonne

Figure 44: Fossil fuel primer

The picture above gives an overview of the main fossil fuels. Even though this book uses the word oil to mean 'oil, the other fossil fuels, and their complex economic and technical relationships', it really focuses on oil itself. This appendix introduces the other fossil fuels, but by no means does it redress the balance.

Conventional natural gas

Methane has long provided energy and hydrocarbons for materials, but it has generally been seen as oil's little sister. Well, that sister has come of age, moved out of her parents' home and got an incredible job. In short, some people see natural gas as the replacement for oil in our economy. It certainly has some credentials in that area: it is abundant in shale formations across the world, it releases more energy per tonne of CO_2 than oil or coal, and we can, relatively, easily remove pollutants from it before we burn it.

It also has its downsides. In addition to the potential for leaks, natural gas requires invasive extraction techniques and still increases levels of greenhouse gases and pollutants in the atmosphere.

It doesn't have a long-term future as an energy source (some experts expect natural gas production to peak this century); however, many advocates believe its near-term advantages make it a potential bridge to a low-carbon future.

Coal

Coal fuelled the Industrial Revolution. It powered our 'dark, satanic mills' and the ships that carried goods across the empires of the 19th century. It continues to be the fuel of choice for electricity generation, despite being the least energy-dense and most polluting of the fossil fuels. Because it does the same jobs as oil and natural gas, its fate is closely tied to theirs.

Until October 2008 US coal consumption stayed fairly steady. Then fracking dropped the price of natural gas, and coal usage fell. But each time natural gas prices increase, electricity producers turn back to coal.

US coal consumption vs natural gas prices for electricity generation

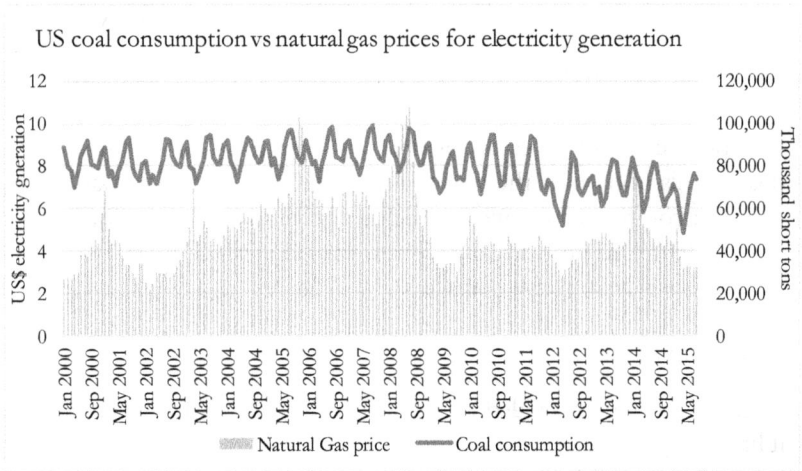

Figure 45: US coal consumption vs natural gas prices

In the Nineties, the UK dashed for gas. In true dash style, the race was over quickly, and coal now holds its own, generating between 25 and 30% of our electricity each year. The next chart tells the story:

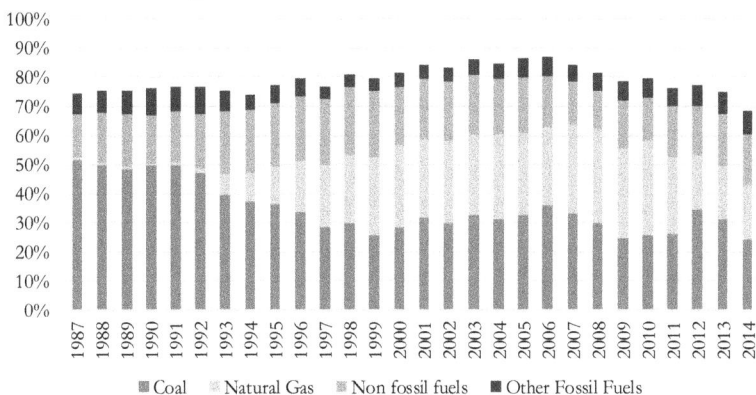

The UK's dash for gas

Coal ▪ Natural Gas ▪ Non fossil fuels ▪ Other Fossil Fuels

Figure 46: The UK's dash for gas

The UK has now declared we will stop using coal for power generation; nevertheless, the future of coal will be bright for some time. Globally we have a lot of it, and the industry is digging new coalfields in Indonesia and Myanmar and continues extraction in Australia and America. Environmental concerns have spurred innovation, and as we saw in "Where has all the oil gone?" carbon capture and storage (CCS) promises the power of coal without the pollution. However, CCS comes at a high price, and so we might assume that the future of this technology is dependent on high oil prices.[227]

From both availability and affordability perspectives, coal appears to offer a grand alternative to oil, so we must consider the downsides. As well as producing more CO_2 for each unit of energy than natural gas or even oil, coal holds other chemicals such as mercury, sulphur and nitrogen. Between them, the compounds of these elements and other pollutants cause over 1,600 premature deaths in the UK, and 22,000 across the whole of the EU, every year.

What's more, modern coal mining doesn't resemble the old pit head. In the UK we are still accustomed to the idea of deep coal mining, with deep shafts and coal faces worked by men and machines.[228] Did you

[227] Even though there is no formal tie between the price of oil and that of natural gas, they respond in similar ways. During the oil price crash of 2014-2016, the price of natural gas dropped 48%.

[228] This is despite there no longer being any deep coal mines in the UK.

know that two-thirds of the coal we burn in Britain comes from surface (not deep) mines? America is the second-largest supplier of coal to the UK, and 70% of its coal comes from surface mining and mountain blowing, often in areas of natural beauty and ecological diversity. Search the Internet for images of 'mountaintop removal' to understand more. An Appalachian, environmental, non-profit organisation estimated that mining has destroyed 500 mountains in that region and Duke University estimates that the central range is 40% flatter than when mining became extensive. Coal might be financially cheap, but we have to ask whether we want to pay the health and environmental costs.

Other sources of hydrocarbons

We have discovered that there are many grades of oil. They start with light, sweet oil and work through to heavy, sour oil, and we need a range of technologies to extract them. There are also several grades of coal, from peat (which has the potential to become coal) to lignite and bituminous coals, to anthracite. So it should be no surprise that hydrocarbon gases also have many forms and sources.

In the appendix "The chemistry of carbon" we learn about alkanes. We use the four lightest alkanes (methane, ethane, propane and butane) as fuel. The heavier ones are liquid and take the familiar name Liquid Petroleum Gas or LPG. The oil and gas industry also calls them natural gas liquids, wet natural gas or condensate. Methane on its lonesome is dry natural gas. Wet and dry natural gases come both from their dedicated wells and from each other's, and both are extracted with oil.

Licence holders in many countries extract the methane adsorbed[229] by coal via a well. Like many terms we have come across, the name for such gas is self-descriptive: Coal Bed Methane (CBM for short).

We could also go back to using gasification to turn coal into (town) gas. But instead of gas works springing up all over the place, Underground Coal Gasification can turn coal into gas underground and pump it out. There are commercial plans to use this process in the UK.

---

229 Sorry, you did not catch a typo. Adsorption occurs when liquids and gases (and dissolved solids) stick to the surface of another chemical. In contrast, absorption happens when a liquid, gas or solid permeates or dissolves in another chemical.

A potted history of Underground Coal Gasification

Sir William Siemens (brother of Werner, whom we met twice in "Taking all my energy") first thought of this method of producing gas in a familiar decade – the 1860s. As with other parts of our tale, the history of this subject is crowded with the efforts, inventions, trials and errors of many people. In this case they ranged from the inventor of the periodic table, Dimitri I Mendeleev, to the discoverer of Argon and the other noble gases, Nobel-winning chemist Sir William Ramsey. However, the credit for developing the forerunners of the methods used today goes to American AG Betts, who registered three patents in 1910. Nonetheless, Sir Ramsey made the first working demonstration two years later at Hett Hill, County Durham.

Another source of natural gas is methane hydrate (also known as methane clathrate), a chemical found in the seabed. The Japanese have successfully extracted small amounts of methane from methane hydrate from below the Pacific, and the (US) National Energy Technology Lab with Conoco Phillips have done the same in Alaska. The technology is still in its early stages. In those two examples the gas took years – and millions of dollars – to produce. To release industrial volumes of gas will take many more years and much more money.

Let's consider how to transport natural gas if you don't have a pipeline. As it is a gas, it takes up a lot of space. There are two ways to overcome this problem. If you super-cool the gas, it turns to liquid natural gas (LNG). Unfortunately, you have to go as low as -170°C. If you squeeze the gas using pressures around 100 times higher than your car tyre pressures, you get compressed natural gas (CNG). We can carry LNG and CNG in tankers and store them in tanks and bottles.

In "Here, there and everywhere" we saw ways of making syngas and biogas from a range of waste products and with microbes. These are not new ideas; the UK already makes the equivalent of 26GWh of gas and hydrocarbons from industrial sources (though not with microbes):

Blast furnace gas and basic oxygen steel furnace (BOS) gas. Made during iron and steel manufacture. They are used to produce electricity and heat, and as energy in industrial processes.

Coke oven gas. Produced when coal is heated to make coke. Coke oven gas is used to produce electricity and heat, and as energy in industrial processes.

Synthetic coke oven gas. A mix of natural gas and blast furnace gas or BOS gas, this is similar to coke oven gas and used for the same purposes.

Benzole. A liquid chemical used to make styrenes and phenols. Styrenes are used to make plastics and rubbers. Phenols are often used as disinfectants and antiseptics, and to make other chemicals.

Tars. By-products of the coke and iron making processes (though historically made by distilling coal) and used as a feedstock or energy.

Because they are abundant and cheap, it seems likely that we will burn methane and its sister fuels for some time to come. They also release far less CO_2 than coal and burn cleaner than both oil and coal. However, we know that we can't burn all the hydrocarbons without irreparable damage to the environment, and have seen that it could take decades for us to convert to natural gas – perhaps just in time to see it peak. You might well ask: why not switch straight to a lower-carbon world?

Appendix Eight – A quick introduction to the UK electricity market

When they evolved, the energy markets met prevailing conditions. Historically, those circumstances featured plenty of cheap fossil fuels and governments willing to pay for security of supply. Every country had these conditions, but different cultures, politics and access to money and energy led to a vast array of energy markets. Here we consider the UK's market for electricity.

The UK electricity market has a long and prestigious history. Godalming had the first electricity market. A hydroelectric station produced the electricity and two overhead circuits delivered it to streetlights, the local mill and a domestic customer. The mill sat close to the power station, and the circuit needed resistors to direct electricity to the town to keep its lights shining brightly.

The same company then produced and delivered Godalming's electricity; the 600 other electricity companies set up by 1923 followed a similar model but produced different electricity. They had different voltages, currents and frequencies, and they supplied to a limited area. This multiplicity of types of electricity persisted until an Act of Parliament (the Electricity (Supply) Act 1926) put in place a body to create a national grid working at 132 kilovolts (kV). The first section of the UK's national grid opened in Scotland in 1930.

In 1947 electricity fell within the scope of nationalisation (via the Electricity Act 1947). In 1957 the industry started to take the shape familiar to us now (via the Electricity Act 1957). The Central Electricity Generating Board produced the electricity. It managed transmission and supplied 12 regional boards, which distributed the electricity and managed retail sales. The Electricity Council consolidated common tasks. Privatisation (via the Electricity Act 1989) kept the basic arrangement and introduced competition. Now licence holders produce electricity and supply it into the grid (competitively). Electricity Supply companies buy that electricity and provide it to us (competitively). Transmission and distribution remain monopolies but are profit-making ventures.

Our electricity bills charge for electricity and connection to the grid. We used to pay for connection with a standing charge, but for most people the unit price (kWh) of electricity now covers it. Your supplier then pays (directly and indirectly):

- Licenced generators;

- National Grid – the System Operator, which balances supply and demand;

- Transmission Operators, which own and maintain the high-voltage transmission equipment (National Grid in England and Wales, Scottish Power in southern Scotland and Scottish Hydro Electric for northern Scotland and the Scottish Islands);

- Distribution Network Operators (DNOs), which transform high transmission voltages and distribute low-voltage electricity in their region (in Great Britain we have fourteen DNOs). Northern Ireland works with the Republic of Ireland to supply electricity.

Ofgem oversees the whole arrangement. In their words:

> *"Ofgem is the Office of Gas and Electricity Markets. We are a non-ministerial government department and an independent National Regulatory Authority, recognised by EU Directives. Our principal duty is to protect the interests of consumers, both present and future. We do this in a variety of ways including:*
>
> *promoting value for money*
>
> *promoting security of supply and sustainability, for present and future generations of consumers, domestic and industrial users*
>
> *the supervision and development of markets and competition*
>
> *regulation and the delivery of government schemes.*
>
> *We work effectively with, but are independent of, government, the energy industry and other stakeholders within a legal framework determined by the UK government and the European Union."*

However powerful and well-intentioned the regulator is, the current system gives the market a strong influence – for example, when the price of fossil fuels increases, the price of 'green' tariffs also goes up. Cynically you could argue that this is just the energy companies trying to profit. Unfortunately, the reality is that suppliers who bid for their electricity drive up the price of renewable energy until it is no longer cheaper than fossil-fuel electricity.

However, the structure of the energy markets is changing rapidly. In 2014 the UK saw (at least) three exciting developments. The first was government-led and changed the way that energy companies contract to supply electricity to the grid. Now generation companies are paid for electricity supplied *and* for committing to maintaining a level of capacity. You might expect this to increase prices, but the government believes it will help keep prices in check; this is because capacity payments give generators an incentive to invest in new facilities and maintain existing stock.

But let's be honest: that is the kind of geeky change that gives the subject of energy a bad name. Much more exciting is the idea of a local government *generating* electricity. As much fun as it is to watch you roll your eyes and moan about a return to socialism and nationalised industries, I feel obliged to remind you that these schemes are no such thing. We met two of them in "Taking oil my energy", and there are many more.

The third change is about the 'big six' – or at least about E.ON. In 2014 the German energy giant announced a strategic move to separate into two companies. E.ON will focus on renewables and nuclear, distribution networks, and customer solutions. Uniper will specialise in power generation from fossil fuels, global energy trading, and exploration and production. The latter started trading on 1 January 2016. Another German energy company, RWE (npower in the UK), made a similar announcement in 2015.

It is impossible to end this appendix without mentioning the findings of the Competition and Markets Authority (CMA). In July 2014 Ofgem asked the CMA to investigate the energy market and determine whether it was uncompetitive. The CMA spent a year studying the market and reported in July 2015 that the market had been skewed by government regulation and a lack of willingness to switch on behalf of customers. It calculated that the UK's 19m dual-fuel customers could save an average of £160 per year each if they switched to the cheapest tariff. What are you waiting for?

Appendix Nine – Risk

Life is a risky business. When we make any decision, we weigh up the likelihood of something bad happening (probability) and the consequence of that negative event (impact). Risk management is the art and science of assessing and improving plans and situations to make them more robust. Please note that the excellent risk people think about threats (bad things) *and* opportunities (good things) when they make a risk assessment.

Say you walk into a room and your favourite treat is sitting on a table. There is no one around and you edge a little closer. Taking a bite, a small handful or a little sip is tempting – so very tempting – but you might be seen. Or be found out. How likely is someone to notice? If your treat is a big, fat, gooey chocolate cake, the hole you leave will be obvious. Not to mention the chocolate all over your face. A single sweet from a haphazardly filled bowl may not be spotted. A sip of wine, beer or orange juice may go unnoticed. The probability of detection is different in each case.

You might think: *So what if someone finds out – what's the worst that can happen?* The consequences of detection also vary. You may have taken a chunk from the favourite cake of a gangland mobster right before his surprise party starts, or you could be in the home of the most generous host on the planet, who insists you take more.

You might really want the treat and realise that, to avoid being caught, you could take a sip and top up that drink, or swish the sweets around to hide the gap. If you are found out, you could blame the cake-snaffling on a racoon that ran into the room (after you wash your face). Would you help yourself?

When it comes to oil, the questions are different (and a lot more serious). We could say there's a risk that we cannot continue to depend on oil to allow the richest billion people to keep their modern, comfortable, safe lives – let alone support the aspirations of the other six billion people. Just like any other risk, we consider probability (let's call that 'how much can we depend on oil?'), consequence ('what might happen?') and treatment ('what can I do differently to avoid the risk or make it less of a problem?').

One politician has suggested the need to 'get an insurance policy' against climate change. He doesn't say that he believes in Climate Change

– just that he doesn't like the consequences if greenhouse gases *are* changing our environment. The politician in question is George Shultz. If you are old enough to remember Ronald Reagan, then you may remember Shultz. No synopsis can convey the universal respect he garnered during his tenure as Secretary of State. Similarly, a short paragraph cannot express the extent of his Republican credentials. Yet he drives an electric car, has domestic solar panels, and is an advocate for action on climate change. In doing so, he appears to have made a risk assessment and decided that it's better to be safe than sorry.

Sources

There are over 1,200 referenced sources for *The VFUU Price of Oil*. Adding them to the book would push its size (and price) to the limits. So all sources are given on the author's webpage.

www.michellespaul.wordpress.com/sources